R. S. Harwood

Marine Physics

Marine Physics

R. E. CRAIG
The Marine Laboratory, Aberdeen, Scotland

1973

ACADEMIC PRESS · LONDON AND NEW YORK

ACADEMIC PRESS INC. (LONDON) LTD.
24/28 Oval Road,
London NW1

United States Edition published by
ACADEMIC PRESS INC.
111 Fifth Avenue
New York, New York 10003

Copyright © 1973 by
ACADEMIC PRESS INC. (LONDON) LTD.

All Rights Reserved
No part of this book may be reproduced in any form by photostat, microfilm, or any other means, without written permission from the publishers

Library of Congress Catalog Card Number: 72–12719
ISBN: 0–12–195050–6

PRINTED IN GREAT BRITAIN BY
ROYSTAN PRINTERS LIMITED
Spencer Court, 7 Chalcot Road
London NW1

Preface

People with many disciplines are engaged in study of the sea and each discipline needs access to the basic premises of the others. I hope that parts of this book are readable enough to be of interest to engineers and biologists who need a little more background in the physical aspects of the sea. I am conscious that the book is not uniformly easy. My treatment of tides and of optical and acoustic questions is basic and practical, while the topic of wind-driven currents is perhaps too theoretical for most readers. The difficult question of surface waves cannot be properly treated at this level or, indeed, by this author. The different chapters do not depend greatly on each other; each is an essay on its own topic.

October 1972   *R. E. Craig*

Contents

Preface	v
1. Density currents in the sea	1
2. Diffusion processes in the sea	9
3. Wind currents in deep water	17
4. Qualitative physical oceanography	36
5. Waves in deep and shallow water	44
6. The tides	55
7. Optics	63
8. Acoustics	73
Appendix: Some physical properties of seawater	81
Index	83

Chapter 1: Density Currents in the Sea

Movements of the sea can be ascribed to several causes. These movements acquire a special character as a result of the rotation of the earth. The prime movers are tidal forces, wind stress, and horizontal differences in density maintained by climatic factors. We concern ourselves first with the last named.

For all practical purposes the density of sea water depends only on its salinity, its temperature and the pressure acting upon it. The appendices give an indication of the relationship between these variables.

If two contiguous water masses have identical relationships between the depth z and the density ρ, then the pressure distribution in each is given by

$$p_z = \int_0^z g\rho \, dz$$

and is of course identical for each mass. Hence there are no unbalanced forces. If the vertical distribution of density is not the same in the two water masses, there will be an imbalance in the pressure field which will induce motion.

Rather than build our theory directly upon the density, it is more convenient to use as a parameter the specific volume, denoted by A, where $A = 1/\rho$. This is of course the volume of unit mass of the water. We had observed that

$$dp = g\rho \, dz \qquad \text{thus} \qquad dz = A/g \, dp$$

and the depth we must proceed below the surface to reach some specified pressure p is evidently given by

$$z_p = \frac{1}{g} \int_0^p A \, dp.$$

Thus if we know A as a function of pressure we can determine the depth of any particular isobaric surface.

(The most useful unit of pressure is the decibar, which is quite close to the pressure at 1 metre depth in sea water. If we assume uniform value of the specific volume we have

$$z_p = \frac{A}{g} \cdot p;$$

that is for one unit of pressure $z_1 = A/g$ where A is typically about 9.75×10^{-4} m^3 kg^{-1} and the unit of pressure is N m^{-2} and g is approximately 9.81 m sec^{-2}. Now

$$1 \text{ decibar} = 10^4 \text{ N m}^{-2}$$

so

$$z(1 \text{ decibar}) = 10^4 \, A/g$$

which is

$$10^4 \times 9.75 \times 10^{-4} \times 1/g = \frac{975}{981} \text{ m}.$$

Thus the depth to a pressure of x db is always in round figures x metres, and we can conveniently speak, for instance, of "the topography of the 200 decibar surface".)

Let us now consider an inhomogeneous mass of water in which the specific volume varies from point to point. It will be clear that since the isobaric surfaces are not parallel to each other, they cannot all be parallel to level surfaces. (By a level surface we mean a gravitational equipotential surface). The effect of the gravitational field will be, in simple verbal terms, to induce flow down the slope of the isobaric surfaces (in the absence of rotation). To be more precise we may say that if the angle at any point between an isobaric surface and a level surface is μ, then the water at that point is subject to a force of $g \sin \mu$ per unit mass. Since we are dealing with small angles, this force is effectively $g\mu$ (μ in radians).

The field of $g\mu$ therefore represents the instantaneous acceleration field of the water particles.

THE ROTATION OF THE EARTH

The length of a sidereal day is 23 hr 56 min 4·09 sec, or 86,164·09 sec, so the angular velocity of the earth's rotation from west to east is $2\pi/86{,}164{\cdot}09$ or $0{\cdot}000073$ rad sec^{-1} in round figures. Now, we wish to use the rotating earth as our frame of reference and it may be shown (e.g. Rutherford,

1939) that the movement of masses in such a frame of reference may be accounted for if we introduce two fictitious forces acting on any particle.

The first of these is the *centrifugal force* $mr\omega^2$ acting perpendicularly away from the axis. We do not need to concern ourselves with this, as it has the effect merely of modifying the gravitational field of the earth. We may assume that the surface of an ocean of uniform density would set itself everywhere perpendicular to the resultant field, giving the familiar figure of the earth, the equatorial radius exceeding the polar radius by 0·3408%.

The second fictitious force, named after Gaspard de Coriolis, is of fundamental importance in oceanography. It is zero for particles which are stationary in the frame of reference, but takes non-zero values for particles which are moving with respect to the frame of reference. The value of the Coriolis force is the vector product $-2m\boldsymbol{\omega} \times \mathbf{v}$ where $\boldsymbol{\omega}$ is the vector rotation and $\mathbf{v}$ is the particle velocity. The magnitude of this product is $2m\omega v \sin \theta$ where θ is the angle between the axis of rotation and the velocity. The direction is perpendicular both to the axis and to the velocity.

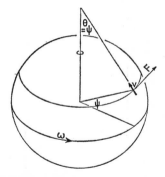

Fig. 1. Coriolis force for the north component of velocity.

To find the horizontal component of this force it is convenient to consider the north and east components of the velocity separately. For the north component of the velocity v_n the Coriolis force is purely horizontal (Fig. 1). We note that θ is for this component equal to the latitude ψ and so the associated Coriolis force is $2m\omega v_n \sin \psi$. This acts to the right in the northern hemisphere, as will be obvious from the consideration that a particle moving poleward is moving into a region of lesser easterly velocity. By a corresponding argument the force acts to the left in the southern hemisphere.

For the east component of velocity, the angle θ is always a right angle, so the Coriolis force is always $2m\omega v_e$, being directed at right angles to the axis of rotation and so being horizontal at the poles and vertical at the equator (Fig. 2a). Clearly the horizontal component of the force is $2m\omega v_e \sin \psi$ and is directed to the right in the northern hemisphere and to the left in the southern. Thus we may sum the effects of the northward and easterly motion giving the horizontal Coriolis force due to horizontal motion in the form $F = 2m\omega v \sin \psi$ (See Fig. 2b).

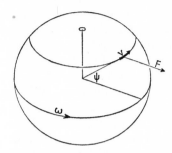

Fig. 2(a) Coriolis force for east component of velocity.

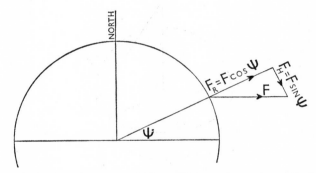

Fig. 2(b) Resolution of Coriolis force for east component of velocity.

Vertical motion of a water mass similarly induces a horizontal force. Here the angle is $(90° - \psi)$ and so the magnitude of the force is $2m\omega v_r \cos \psi$. These important aspects of the Coriolis force are set out in Table 1.

Table 1. The horizontal Coriolis forces

Motion of particle	Northern hemisphere	Southern hemisphere
Horizontal	$2m\omega v \sin \psi$ to right	$2m\omega v \sin \psi$ to left
Upward	$2m\omega v \cos \psi$ to west	$2m\omega v \cos \psi$ to west
Downward	$2m\omega v \cos \psi$ to east	$2m\omega v \cos \psi$ to east

CLASSICAL DYNAMICS

The proposition that a set of observations represents a steady state condition has obvious approximate validity if we are considering a geographically large enough part of a sea or ocean. If repetition of observations discloses a consistent density pattern, we can draw up charts showing the topography of the various isobaric surfaces. In the absence of rotation, these would imply an acceleration field, and the state could not be a steady one. It can however be a steady state on the rotating earth if the accelerations due to density are everywhere balanced by the Coriolis forces due to movement. Thus if we have a value μ for the angle at any point between an isobaric and a level surface, this implies an acceleration μg as explained earlier. If the water has a motion v at right angles to the slope, the Coriolis force may exactly cancel the effect of the slope. Only if this is so can we justify the observed condition of a steady state. The velocity v gives rise to a Coriolis acceleration $2\omega v \sin \psi$, so the required condition is

$$2\omega v \sin \psi = \mu g$$

or

$$v = \mu g / 2\omega \sin \psi.$$

Rather than derive a universal formula let us suppose $\mu = 10^{-5}$ or 1 cm km^{-1}, and the latitude ψ to be $30°$ then

$$v = 10^{-5} \times 9\cdot 81/(2 \times 7\cdot 3 \times 10^{-5} \times 0\cdot 5) = 9\cdot 81/7\cdot 3 = 1\cdot 3 \text{ m sec}^{-1}.$$

This is about 2·6 knots, and would of course be a very rapid water movement by oceanographic standards.

VERTICAL COMPONENT OF CORIOLIS FORCE

The vertical component of Coriolis acceleration is due entirely to the east-west component of water velocity. Let us suppose in the previous example that the flow was directly to the east at $1\cdot 3 \text{ m sec}^{-1}$. Then the vertical component of acceleration is

$$2\omega v_e \cos \psi = 2 \times 7\cdot 3 \times 10^{-5} \times 1\cdot 3 \times \cos \psi = 1\cdot 9 \times 10^{-4} \times \cos \psi$$

at, for example, $30°$ north or south latitude this represents an upward acceleration of

$$1\cdot 9 \times 10^{-4} \times 0\cdot 87 = 1\cdot 7 \times 10^{-4} \text{ m sec}^{-2}$$

or $0\cdot 000017\, g$ that is to say the "apparent density" is reduced by $0\cdot 0017\%$ which, even though the velocity concerned is high, is beyond the limits of accuracy with which density can be determined by oceanographic methods. For practical purposes therefore we may regard the vertical component of the Coriolis force as negligible.

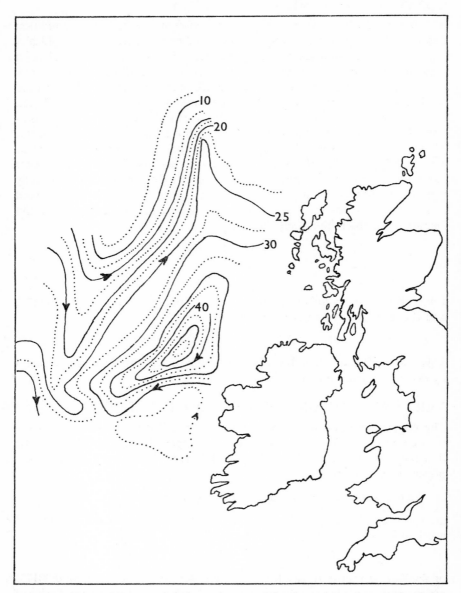

Fig. 3. Height of the sea surface (in centimetres, zero arbitrary) measured from the 1500 decibar surface. This was in June 1951 and is based on 62 hydrographic stations. (After Tait and Tulloch, 1959).

1. DENSITY CURRENTS IN THE SEA

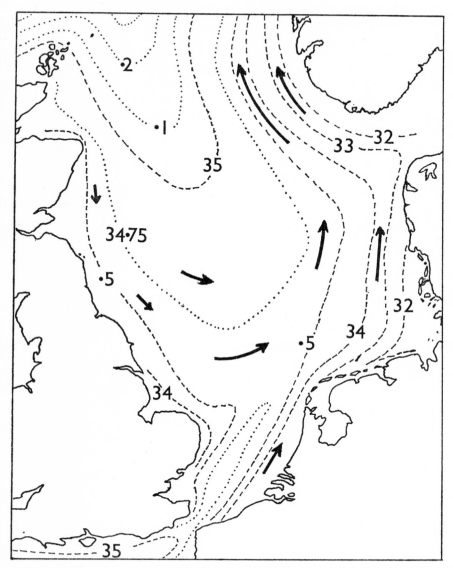

Fig. 4. Mean surface salinity of the North Sea (from I.C.E.S. Atlas) and dominant surface water movements.

LIMITATIONS

The fundamental difficulty in applying these relationships to the determination of water movement, is that there is no means of directly recognising a level surface at sea. Thus no absolute reference is available and the most that can be done is to determine the *relative* velocity of the water at different levels. If, for example, it is found that at some specified level, or below some specified level, water movement is effectively zero, the implication is that the isobaric surfaces at this depth are parallel to level surfaces. If any such assumption can be made, the absolute velocity of the upper layers can be estimated by the foregoing methods. However any such assumption invites a direct check, and methods capable of measuring the deep currents directly can in general be applied equally to measuring the upper water currents also. Hence it is of doubtful utility to attempt to specify the current regime of an area purely by observations of density. The most that might be claimed is that these classical methods provide a framework which aids the understanding of oceanographic processes and helps to indicate the crucial regions for more direct studies.

An example of such observations, and their results is given in Fig. 3, which is taken from Tait and Tulloch (1959), and shows the situation at one particular time in a region to the west of the British Isles. The theory also explains the gross features of coastal currents. In the northern hemisphere, wherever there is significant dilution of the sea by rivers, the most usual coastal water movement is to the right of an observer looking seaward. One might travesty the more exact rules of electromagnetics by giving the dynamic rule in the elementary form: "Look towards the denser water. The current in the northern hemisphere will be to your right".

As a further example we might note that density in the North Sea is dominated by the dilution process. Figure 4 shows the mean salinities and the gross circulation which can be deduced from it. The general features of this circulation are supported by other observations. To determine and understand the details and day-to-day variations in this pattern is a task as yet only beginning.

BIBLIOGRAPHY

Atlas de Temperature et Salinité de l'Eau De Surface de la Mer du Nord et de la Manche (1933). *Cons. Int. Explor. Mer. Copenhagen.*
Proudman, J. (1953). "Dynamical Oceanography". Methuen, London.
Rutherford, D. E. (1939). "Vector Methods", 127 pp. Oliver and Boyd, London.
Tait, J. B. and Tulloch, D. S. (1959). Hydrography of the North-Western approaches to the British Isles. *Mar. Res.* **1,** 32.

Chapter 2: Diffusion Processes in the Sea

The transfer of properties between adjacent regions in the sea can take place by translation or by turbulent diffusion. (Ordinary molecular diffusion is orders of magnitude less than turbulent diffusion and need not be here considered).

Turbulent diffusion results from the existence of a spectrum of eddies which transfer properties within the water mass, even in the absence of an overall resultant motion. It will be evident that any mixing process of this kind could be examined on a microscopic scale and seen as a matter of transport by water movement, or looked at on a larger space and time scale as a matter of diffusion.

To take a simple example of effluent being discharged from a pipe into a tideway, the result is a plume stretching away in the direction of the local tidal current. The increasing breadth of the plume depends in a simple manner on the horizontal diffusion coefficient associated with the water mass. If however we have to study the distribution of this effluent to great distances, and over a period of months, we come to regard the to-and-fro, or elliptical tidal flow as merely a component of the diffusion process. In general, any particular investigation will involve a choice of the scale of motion to be treated as turbulent, and above which the motion will be regarded as translational. The choice is arbitrary, and will depend on the nature of the investigation.

To put the subject upon a quantitative basis, we consider a dissolved substance whose concentration at any point is $c \, \text{kg m}^{-3}$. The gradient of concentration in the x direction may be written as dc/dx and the rate of transfer of the substance, per m^2 across a surface perpendicular to the x axis, is then $- K(dc/dx) \, \text{kg sec}^{-1}$ and this expression defines the coefficient of diffusion K which has the dimensions $m^2 \, \text{sec}^{-1}$.

Diffusion in a horizontal direction takes place much more readily than in a vertical direction, since in general a water column shows a certain degree of stability (density increasing with depth) and this stability is a drastic inhibitor of vertical turbulence and hence of vertical diffusion. As might be expected, wave action and shear velocities are notable stimulants of turbulence.

The observed values of the horizontal diffusion coefficient are so variable that "typical" values are not easy too quote, however open sea values are generally in excess of 10 m² sec⁻¹ and values as high as 10^3 are necessary to account for the diffusion observed in well mixed tidal estuaries. Such high values apply in the longtitudinal direction in estuaries with tidal rates of around $1 \cdot 5$ m sec⁻¹, and refer to long term processes: i.e. those in which the tidal action is regarded as part of the turbulent process.

The vertical coefficient of diffusion is in general some orders of magnitude less than the horizontal, values in the range 10^{-3} to 10^{-5} being quoted by Gade (1968) for parts of the Oslofiord. Whenever a marked density discontinuity occurs either on account of a jump in temperature or salinity, the values of K (vertical) are observed to be even less than are above quoted. Thus for the purpose of some pollution studies for example, the density discontinuity acts in a similar manner to the sea bed, in that it effectively inhibits downward diffusion. Thus we can treat some dispersion problems as 2-dimensional, thereby allowing them to be treated in a simple way.

We may, as an example, consider the spread of a patch of pollutant or of a dye for oceanographic studies. Suppose this is introduced into the sea, and the depth to the seabed or discontinuity is z metres. We shall suppose that the mass of pollutant is m and that the diffusion coefficient K is the same in the x and y directions.

We shall show that (a) the form of the patch will approximate to a bivariate normal distribution and (b) that the effective area of the patch increases proportionately to the time and proportionately also to K.

If we write for the concentration

$$c = \frac{m}{z\alpha} \exp\left(-\frac{\pi r^2}{\alpha}\right)$$

where r is the distance from the centre of the patch, then we may check by integration that this represents a total mass m of the pollutant, for

$$\int_0^\infty 2\pi r z c \, dr = m.$$

The centre value (at $r = 0$) is $c = (m/z\alpha)$, thus the quantity of pollutant is such that the mass m would fill a volume $z\alpha$ at this concentration. Also we note that the concentration has fallen to $1/e$ of this centre value at a radius r such that $\pi r^2 = \alpha$.

The mass of pollutant present outside the limit $\pi r^2 = \alpha$ is

$$m\left[-\exp\left(-\frac{\pi r^2}{\alpha}\right)\right]_{\sqrt{\alpha/\pi}}^\infty = \frac{m}{e}.$$

2. DIFFUSION PROCESSES IN THE SEA

We now have to show that a distribution of this form is stable in respect to diffusion, that is, that the effect of diffusion is simply to increase the characteristic area α of the distribution. Our assumed distribution was

$$c = \frac{m}{z\alpha} \exp\left(-\frac{\pi r^2}{\alpha}\right).$$

The outward flow of pollutant through a cylindrical surface at radius r is

$$F = -2\pi Kzr \frac{dc}{dr}.$$

So the rate of accumulation of pollutant in the annulus $(r, r+dr)$ is

$$-\frac{dF}{dr} dr = 2\pi Kz \left(\frac{dc}{dr} + r\frac{d^2c}{dr^2}\right) dr.$$

Now the volume of the annulus is $2\pi zr\, dr$ and so the rate of change of concentration at a radius r is

$$\frac{dc}{dt} = K\left(\frac{1}{r}\frac{dc}{dr} + \frac{d^2c}{dr^2}\right)$$

$$= \frac{\pi m K}{z\alpha}\left(\frac{4\pi r^2}{\alpha^2} - \frac{4}{\alpha}\right) \exp\left(-\frac{\pi r^2}{\alpha}\right)$$

$$= \frac{4\pi K}{\alpha}\left(\frac{\pi r^2}{\alpha} - 1\right)\frac{m}{z\alpha} \exp\left(-\frac{\pi r^2}{\alpha}\right)$$

$$= \frac{4\pi K}{\alpha} c \left(\frac{\pi r^2}{\alpha} - 1\right).$$

We may note this is positive for $\pi r^2 > \alpha$ and negative for $\pi r^2 < \alpha$.

We need to be satisfied that this effect of diffusion on the distribution is identical in form to the effect of changing the value of α. To show this we evaluate

$$\frac{dc}{d\alpha} = \frac{m}{z\alpha^3} \pi r^2 \exp\left(-\frac{\pi r^2}{\alpha}\right) - \frac{m}{z\alpha^2} \exp\left(-\frac{\pi r^2}{\alpha}\right)$$

which has the same form as the previous expression. To obtain the time scale of the spreading we note that

$$\frac{d\alpha}{dt} = \frac{dc}{dt}\frac{d\alpha}{dc} = \frac{4\pi K}{\alpha}\alpha = 4\pi K.$$

Thus
$$d\alpha = 4\pi K dt$$

or
$$\alpha = 4\pi K t$$

if $\alpha = 0$ when $t = 0$.

EDDY CONDUCTIVITY AND EDDY VISCOSITY

These quantities are analogous to eddy diffusion. In the first case we can consider *heat energy* as the dissolved property exchanged by turbulence. In round figures we can consider one cubic metre of sea water as requiring 4.2×10^6 joules (1,000 kilocalories) to change its temperature by 1°C.

By analogy with what has gone before we may write

$$\text{Heat flow (W m}^{-2}) = -K \frac{dc}{dx}$$

where c is an energy concentration in joules per square metre. Now a temperature gradient of one degree per metre represents a concentration gradient of 4.2×10^6 J m^{-4}. Thus if we define the thermal conductivity μ by the relationship

$$\text{Heat flow (W m}^{-2}) = -\mu \frac{dT}{dx}$$

we see that $\mu = K \times 4.2 \times 10^6$ and the dimensions of μ are W m^{-1} deg^{-1}.

In a similar way (but the analogy is not really exact) we may account for viscosity in terms of a turbulent transfer of momentum between adjacent layers in relative motion. For a relative motion of v m sec^{-1} the momentum per cubic metre of sea water is $1,000\,v$ kg m sec^{-1}. Suppose for example that v is directed in the y direction, and that there is a shear, or velocity gradient encountered as we move in the x direction, then

$$\text{Momentum flow in the } x \text{ direction} = -K \frac{dc}{dx} = -1,000\, K \frac{dv}{dx}$$

$$= -\eta \frac{dv}{dx} \text{ (N m}^{-2})$$

and η the coefficient of eddy viscosity has the dimensions kg m^{-1} sec^{-1}.

THE THERMOCLINE AND NUTRIENT CIRCULATION

One example of the way in which these relationships are of interest is the way in which a thermocline or temperature discontinuity is formed at certain seasons. A general look at the seas around Britain shows that the heat gained by the sea during May and June is just about enough to raise the temperature of the upper 50 m by 5°C. This means that averaged over day and night there is absorbtion by the sea surface of about 200 W m^{-2}.

One might expect to find some kind of smooth gradation of temperature from the surface downward, but this does not happen in offshore areas. What does happen is that any such smooth gradation is upset by the action of wind and waves which strongly mix the upper water so that a uniform upper layer is formed. The difference in density between this upper layer and the cooler water below inhibits free mixing, and a discontinuity appears at perhaps 30 m depth, which gradually deepens during the season.

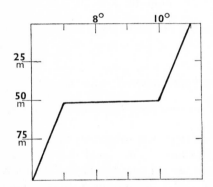

Fig. 1. Idealized vertical temperature structure, early summer.

Thus we might find by observation a temperature distribution, such as in Fig. 1, where in the upper water the temperature decreases by about 1°C in 50 m, with a similar or lesser gradient in the deep water. At the discontinuity we may find a drop of 3°C in 1 m. Successive observations satisfy us that certainly less than a tenth of the heat supplied has reached the deep water beneath the thermocline, and we remember that

$$\text{Heat flow} = -\mu \frac{dT}{dz}$$

hence

$$\mu = -\frac{dz}{dT} \times \text{(heat flow)}$$

This gives for the upper water

$$\mu = 200 \times 50 = 10{,}000$$

and in the discontinuity layer

$$\mu = 20 \times 0\cdot02 = 0\cdot4 \text{ W m}^{-1} \text{ deg}^{-1}$$

From these values of μ we can obtain the corresponding values of K i.e. since $K = \mu/(4\cdot2 \times 10^6)$ we have

for the upper water $K = 2\cdot4 \times 10^{-3} \text{ m}^2 \text{ sec}^{-1}$

for the discontinuity $K = 10^{-7} \text{ m}^2 \text{ sec}^{-1}$.

These small values are not particularly convenient to think about and we shall rewrite them using the day instead of the second as our unit of time, giving

for the upper water $K = 210 \text{ m}^2 \text{ day}^{-1}$

in the discontinuity $K = 0\cdot009 \text{ m}^2 \text{ day}^{-1}$.

Suppose that a nutrient was originally distributed through the water column at a concentration $c \text{ kg m}^{-3}$ and that photosynthesis in the upper 20 m begins to use up ten per cent per day of the nutrient stock: i.e. there is a total uptake of $2c$ kg per day in a 1 m² column. For the sake of simplicity let us suppose a gradient of $0.1c \text{ m}^{-1}$ was set up between 20 and 21 metres, then the flow of nutrient upwards would be

$$210 \times 0\cdot1c = 21c \text{ kg m}^{-2} \text{ day}^{-1}$$

In short, the diffusion process can easily circulate nutrient *within* the water mass over the discontinuity. However to feed this quantity of nutrient up *through* the discontinuity is quite another matter.

Suppose a concentration difference Δc were established across the 1 metre thickness of the discontinuity layer then we can work out the value of Δc required to induce the required upward flow of $2c \text{ kg m}^{-2} \text{ day}^{-1}$. This would be the value given by

$$2c = 0\cdot009 \times \Delta c$$

or
$$\Delta c = 220c.$$

Now the maximum conceivable value of Δc is evidently c which would occur if the upper layer were completely exhausted of nutrient.

Thus using the results of quite elementary observations it is evident that, where such a thermal structure occurs, the plants in the photic zone are able to draw on nutrient available above the discontinuity, but are unable to obtain supplies from beneath the discontinuity until the latter is broken down by storms or other disturbance.

VERTICAL TRANSFER OF STRESS

We next consider the influence of a stratified situation (as shown in Fig. 1) on the nature of horizontal movement that might, for example, be imposed by wind stress. We noted, without making the statement explicitly that the coefficient of eddy viscosity η could in principle be obtained from the coefficient of eddy diffusion K by the relationship $\eta = \rho K$, and this gives us values of the eddy viscosity as follows:

above and below the discontinuity $\eta_1 = 2 \cdot 4 \text{ kg m}^{-1} \text{ sec}^{-1}$

in the discontinuity $\eta_2 = 10^{-4} \text{ kg m}^{-1} \text{ sec}^{-1}$

If we suppose a wind stress to set up a velocity of the upper water of 1 m sec^{-1}, then stresses will be created by the viscosity causing the lower layers of water to move. If we suppose the water stationary at the top of the discontinuity the velocity gradient in the upper layer will be

$$-\frac{dv}{dz} = \frac{1}{50} \text{ sec}^{-1}$$

and the shear stress throughout this layer will be

$$\frac{1}{50} \times 2 \cdot 4 = 0 \cdot 048 \text{ N m}^{-2}.$$

Now suppose the whole of the upper water to be brought into uniform motion, leaving the deep layer stationary, the velocity gradient across the discontinuity will be

$$-\frac{dv}{dz} = 1 \text{ sec}^{-1}$$

and the shearing stress through the discontinuity will be only $10^{-4}\,\mathrm{N\,m^{-2}}$. Thus in the conditions presupposed, the downward transmission of horizontal stress through the 50 m of mixed water is about 500 times more effective than the transmission through 1 m of the very stable discontinuity layer.

It is important to beware of the artificial nature of this example. Water movements in response to wind are dealt with in another section, and are influenced by the earth's rotation, and by boundary conditions. Also it has been found empirically that in conditions even of moderate stability, the coefficient of eddy viscosity in a vertical direction is *not* accurately given by the relation $\eta = \rho K$ but tends to be around an order of magnitude higher. The explanation for this is not offered in an exact way, but it lies in the consideration that in stable conditions, vertical turbulence exchanges various sized masses of water across any horizontal boundary, but owing to the principle of Archimedes there is a tendency for these masses to revert back to their original stratum. It is not implausible to recognise that the exchange of momentum can take place more rapidly than the exchange of such properties as salinity or temperature, which latter requires the break-up and mixing of the transferred water mass with the water across the boundary.

In spite of this qualification, the argument for regarding the water below a sharp temperature or salinity discontinuity as effectively decoupled from the upper water, so far as horizontal stress is concerned, seems to have validity.

BIBLIOGRAPHY

Gade, H. G. (1968). Horizontal and vertical exchanges and diffusion in the water masses of the Oslofiord. *Helgoländer wiss. Meersunters.* **17**, 455–517.

Johnston, R. (1973) "Marine Pollution". Academic Press, London and New York.

Chapter 3: Wind Currents in Deep Water

In this chapter we aim to examine, much more closely than before, the effects of the earth's rotation and the effect of the Coriolis force on the dynamics of wind and water.

The reader will recall that a mass m moving with a velocity u appears to be subject to a force on it equal to $2mu\omega \sin \phi$ where ω is the angular velocity of the earth (35×10^{-6} rad sec^{-1}) and ϕ is the latitude. In the Northern hemisphere this force is directed to the right of the direction of the motion.

Throughout this chapter we shall assume that we are studying systems in 30° *north latitude* so the Coriolis force reduces to $mu\omega$. By this means we can simplify the look of the algebra, while retaining the symbol ω in its proper situation in the equations. It may be helpful to look at certain elementary aspects of dynamics on a rotating system before facing the more specific questions of marine currents. The reader who is expert in dynamics may happily omit the following two sections.

UNDAMPED MOTION OF A MASS AT 30° N LATITUDE

We here suppose a body of mass m situated at the origin, and let a force F be applied to it in the x-direction, continuously from the instant $t = 0$. We shall write $F/m = \alpha$ i.e. α is the acceleration at $t = 0$, and we shall denote differentiation with respect to t by dots.

The equations of motion are:

$$\ddot{x} + \omega \dot{y} = \alpha \qquad (1)$$

$$\ddot{y} - \omega \dot{x} = 0 \qquad (2)$$

from (2)

$$\ddot{x} = \frac{1}{\omega} \dddot{y}$$

and substituting in (1) we have

$$\dddot{y} + \omega^2 \dot{y} = \omega \alpha$$

a particular integral exists in the form $y = \alpha t/\omega$ and the general solution is

$$y = A + B \cos(\omega t + \phi) + \alpha t/\omega$$

and at

$$t = 0, \quad y = 0, \quad \dot{y} = 0 \quad \text{and} \quad \ddot{y} = 0$$

hence

$$\phi = 0, \quad A = \alpha/\omega^2, \quad B = -\alpha/\omega^2$$

and so

$$y = \frac{\alpha t}{\omega} + \frac{\alpha}{\omega}(1 - \cos \omega t).$$

From (2)

$$\dot{x} = \frac{1}{\omega}\ddot{y} \quad \text{so} \quad x = \frac{1}{\omega}\dot{y} + E$$

i.e.

$$x = \frac{\alpha}{\omega^2} + \frac{\alpha}{\omega^2} \sin \omega t + E$$

and $x = 0$ when $t = 0$ so $E = -\alpha/\omega^2$

hence

$$x = \frac{\alpha}{\omega^2} \sin \omega t.$$

As will be seen from Fig. 1 the motion can be described as the sum of two components, a motion in a circle of radius α/ω^2 at a velocity α/ω and a

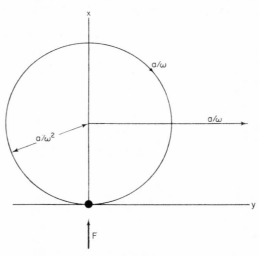

Fig. 1. Component of undamped motion.

uniform motion in the y-direction at the same velocity α/ω. The angular velocity in the circular motion is ω, and the radius is such that the applied acceleration α is just equal to the centripetal acceleration that would be associated normally with that circular motion.

Combining the two motions we see (Fig. 2) that the mass moves in an identical manner to a point on the circumference of a wheel rolling on the y axis. When $t = 2n\pi/\omega$ the mass is again stationary in the y axis. Its maximum velocity of $2\alpha/\omega$ occurs when $t = (2n - 1)\pi/\omega$.

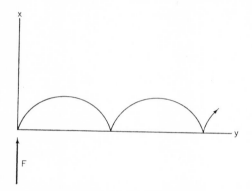

Fig. 2. Resultant undamped motion.

DAMPED MOTION OF A MASS AT 30°N. LATITUDE

Here we suppose the same conditions as above, but we introduce a force opposing and proportional to the velocity. If we suppose this viscous force to be $\dot{x}f$ and $\dot{y}f$ and write for convenience $\xi = f/m$ then the equations of motion become:

$$\ddot{x} + \omega\dot{y} + \xi\dot{x} = \alpha \qquad (1)$$

$$\ddot{y} - \omega\dot{x} + \xi\dot{y} = 0 \qquad (2)$$

from (2)

$$\dot{x} = \frac{1}{\omega}\ddot{y} + \frac{\xi}{\omega}\dot{y}$$

and substituting in (1)

$$\dddot{y} + 2\xi\ddot{y} + (\omega^2 + \xi^2)\dot{y} = \omega\alpha$$

a particular integral exists in the form of

$$y = \frac{\omega\alpha}{\omega^2 + \xi^2}t$$

and the general solution is

$$y = A + Be^{-\xi t} \cos(\omega t + \phi) + \frac{\omega \alpha}{\omega^2 + \xi^2} t$$

that is, by comparison with the previous case, we see that the circular motion dies away exponentially, being reduced to $\exp(-2\pi\xi/\omega)$ after one rotation.

The ultimate velocity in the y direction is

$$\dot{y} = v = \frac{\omega \alpha}{\omega^2 + \xi^2}.$$

Similarly

$$x = A' + Be^{-\xi t}(\cos \omega t + \phi) + \frac{\xi \alpha}{\omega^2 + \xi^2} t$$

and the ultimate velocity in the x direction is

$$\dot{x} = u = \frac{\xi \alpha}{\omega^2 + \xi^2}.$$

Thus the ultimate motion is in a straight line.

The velocity is given by

$$V^2 = u^2 + v^2 = \alpha^2 \frac{\omega^2 + \xi^2}{(\omega^2 + \xi^2)^2}$$

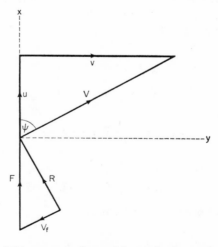

Fig. 3. Ultimate velocity and forces in damped motion.

3. WIND CURRENTS IN DEEP WATER

or
$$V = \alpha(\omega^2 + \xi^2)^{-\frac{1}{2}}.$$

The angle ψ which this motion makes with the applied force is given by

$$\tan \psi = v/u = \omega/\xi.$$

If the mass of the body is m, our symbols imply that there is a viscous force onto it of $Vf = Vm\xi$ which is directed towards the origin in a direction opposite to V (Fig. 3). Now it will be evident that

$$\frac{Vf}{F} = \xi/(\omega^2 + \xi^2)^{\frac{1}{2}} = \frac{u}{v}.$$

Thus, as might have been foreseen, the triangle of forces is similar to the triangle of velocities, and the resultant force R on the body is perpendicular to the resultant motion. Furthermore

$$R = Vm\xi \tan \psi = \omega m V$$

which again is simply in conformity with the original Coriolis law.

In oceanography much use (perhaps too much use) is made of the concept of the steady state, and we deal constantly with this fundamental relationship which is simply that an applied force results in, or is at least directly associated with, a momentum

$$P = mV = F/\omega$$

in the appropriate direction at right angles. Components of force can be regarded as giving rise to corresponding components of momentum.

MOTION WHEN APPLIED FORCE IS REMOVED

If the applied force is removed, or a counter-balancing force applied, which comes to the same thing, then the motion of the body will continue.

In the case of an undamped motion, if the body has a velocity u at the instant when the applied force is removed, then clearly it will continue to move in a circle. The velocity will be unchanged, the circle will be traversed at an angular velocity ω, and so the radius will be given by

$$r\omega = u \quad \text{or} \quad r = u/\omega.$$

With similar initial conditions, but with damping, the radius of the "circle" will initially be $r_0 = u/\omega$ as above, but this radius will decrease and r must be written $r = r_0 e^{-\xi t}$. Thus the motion will be in a spiral and the angle which the velocity at any instant makes with the radius is given by

$$\tan\theta = -\frac{rd\theta}{dr} = \frac{\omega}{\xi}.$$

These conclusions are illustrated in Fig. 4. It may be remarked, with regard to the second example, that the ultimate effect of the frictional forces is to bring the body to rest at the centre of the spiral. However, the effect of the forces is such that the body never moves towards its final resting place. Instead it proceeds with decreasing velocity, but always in a direction inclined to the left by $\tan^{-1}(\omega/\xi)$ from the "direct route home". If $\xi = \omega$ then the angle from direct homing is always 45°.

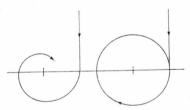

Fig. 4. Motion of undamped and damped particle on removal of applied force.

To take some practical examples, suppose we are dealing with an initial motion at 10 m sec^{-1}. In the undamped condition the free motion will be in a circle of radius

$$\frac{u}{\omega} = \frac{10}{35} \times 10^6 \text{ m},$$

or about 300 km. If the original velocity were $0\cdot1 \text{ m sec}^{-1}$ (1/5 knot) then the motion will be in a circle of radius 3 km.

The case of damped motion leads to conceptually significant results only when we are dealing with large masses. Consider for example a mass of 1 million tons. To obtain the 45° spiral mentioned above we must have

$$\xi = \omega = 35 \times 10^{-6} \text{ sec}^{-1}.$$

Thus the resistive force per 1 m sec^{-1} of velocity must be

$$\xi m = 35 \text{ ton}.$$

A CLOSER LOOK AT THE NATURE OF VISCOSITY FORCES

In the previous chapter we introduced the idea of viscosity, expressed as a diffusion of momentum across a surface, giving rise to a stress per unit area of the surface. The stress, or rate of momentum transfer per unit area, is expressible in the form

$$\text{Stress} = S_x = \mu \frac{du}{dz}$$

for stress in the x direction due to transfer of x momentum in the z direction, and of course

$$S_y = \mu \frac{dv}{dz}$$

for the y components of stress and velocity.

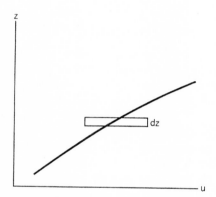

Fig. 5. Force on an element of volume.

For positive values of du/dz and dv/dz the stress acts in the positive direction of x and y on the fluid beneath the surface $z = z_0$ and of course in the opposite direction on fluid above this surface (z positive upwards) as shown in Fig. 5. The force on a volume of unit area lying between z and $z + dz$ is clearly

$$dF_x = \frac{d}{dz}\left(\mu \frac{du}{dz}\right) dz.$$

If μ is independent of z this reduces to

$$dF_x = \mu \frac{d^2 u}{dz^2} dz$$

and so the force per unit volume is simply

$$F_x = \mu \frac{d^2 u}{dz^2}$$

with a similar expression for the y component of velocity. The force is directed in the x-positive direction if d^2u/dz^2 is positive.

The dimensions of μ have been previously noted to be $\text{kg m}^{-1}\text{ sec}^{-1}$ and we previously noted also that the diffusion coefficient associated with momentum transfer was not necessarily the same as that associated with the transfer of chemical properties. For the sake of exactness we specify a *diffusion coefficient for momentum* with the symbol v, putting $v = \mu/\rho$. The dimensions of v are now $\text{m}^2\text{ sec}^{-1}$, as for any other diffusion coefficient; however v is generally referred to as the *kinematic viscosity*.

MOTION OF FLUID IN THE PRESENCE OF VISCOUS FORCES

We continue to assume a situation in 30°N latitude. As the fluid under consideration we initially take the air, and so we use axes in which z is positive upward. Since the ultimate aim is to examine the topic of wind-driven currents, it is evident that we are concerned with horizontal motion and its relation to stresses.

It is exceedingly important to be clear about frames of reference. The argument when presented, will focus attention successively on different reference levels, and will make a point of examining the same basic velocity structure from the point of view of different observers. Thus, we may wish to describe a movement pattern as seen by an observer at high altitude, moving with the upper wind or, at another time, to place our observer at a selected arbitrary level z_0. Again we shall as required use an observer at the air/water interface, or deep down in the sea, partaking of the steady motion at that level.

The differential equations connecting the forces on unit volume of fluid at any level z may be written

$$\omega \rho v + \mu u'' = 0$$

$$-\omega \rho u + \mu v'' = 0$$

where the dashes indicate differentiation with respect to z. These may be reduced to the simpler form

$$\omega v + v u'' = 0 \tag{1}$$

$$-\omega u + v v'' = 0 \tag{2}$$

3. WIND CURRENTS IN DEEP WATER

from (2) we may obtain

$$u'' = \frac{v}{\omega} v''''$$

and substituting in (1) we have

$$v'''' + \frac{\omega^2}{v^2} v = 0$$

which leads to the auxiliary equation

$$D^4 + \omega^2/v^2 = 0$$

whence

$$D = \pm \sqrt{\frac{\omega}{v}} \cdot \frac{1+i}{\sqrt{2}} \quad \text{or} \quad \pm \sqrt{\frac{\omega}{v}} \frac{1-i}{\sqrt{2}}.$$

Let

$$q = \sqrt{\frac{\omega}{2v}}$$

then

$$v = A e^{-qz} \sin(qz + \phi) + B e^{-qz} \sin(qz + \theta)$$

At this stage we must determine our frame of reference. Let us suppose the observer to be situated in the upper air, at $z = \infty$, and further that the motion at an arbitrary level $z = 0$ is along the x axis. Thus at $z = \infty$,

$$u = v = 0$$

and at $z = 0$,

$$u = u_0, \quad v = 0.$$

Our expression for v with these boundary conditions becomes

$$v = B e^{-qz} \sin qz$$

and

$$u = (v/\omega) v'' = \frac{1}{2q^2} v''$$
$$= -B e^{-qz} \cos qz,$$

but since $u = u_0$ at $z = 0$ we have $B = -u_0$ and the solutions become

$$u = u_0 e^{-qz} \cos qz \tag{3}$$

$$v = -u_0 e^{-qz} \sin qz. \tag{4}$$

This result is illustrated in Fig. 6(a) from the point of view of an observer in the upper air, and in (b) from the point of view of an observer moving

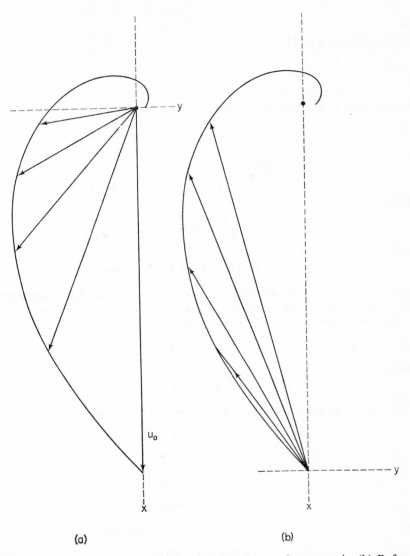

Fig. 6. Spiral of velocities. (a) Referred to an observer in upper air. (b) Referred to an observer moving at μ_0.

at the uniform velocity u_0 with the air at the arbitrary level $z = 0$. Since the exponent and the angular term have the same value qz in the above equations, we recognise that in plan the heads of the vectors lie in a spiral, and that the angle which this spiral makes with any radius is 45°.

Thus for example from Fig. 6(b) the observer at $z = 0$ sees the upper air moving at a velocity $-u_0$. The air immediately above him is however moving, though slowly, at an angle of 45° relative to $-u_0$. Thus the stress at *any* arbitrary level is always directed at 45° to the left of the ultimate velocity of the upper air.

The stress at $z = 0$ can be deduced by reference back to equations (3) and (4) from which

$$u' = -qu_0 \, e^{-qz} (\cos qz + \sin qz)$$
$$v = -qu_0 \, e^{-qz} (\cos qz - \sin qz)$$

i.e. at $z = 0$

$$u_0' = v_0' = -qu_0$$

so

$$V_0' = \sqrt{(u_0'^2 + v_0'^2)} = \sqrt{2}\, qu_0 = u_0 \sqrt{\frac{\omega}{\nu}},$$

thus the stress at $z = 0$ is

$$S = \mu u_0 \sqrt{\frac{\omega}{\nu}} = u_0 \sqrt{(\omega \rho \nu)}$$

since $\mu = \rho \nu$.

If we choose to let z_0 be sea level, then $-u_0$ becomes the high level (undisturbed or geostrophic) wind which we can write as U and we then have the fundamental result that the magnitude of the stress at the sea surface is

$$S = U\sqrt{(\omega \mu \rho)}$$

where μ and ρ refer to the air, and the stress is at 45° to the left of the geostrophic wind.

VOLUME TRANSPORT AND MOMENTUM CHANGES

Referred to the upper air the volume transport across a vertical surface of 1 m lateral extent in the appropriate direction is

$$T_x = \int_0^\infty u \, dz = U \int_0^\infty e^{-qz} \cos qz \, dz = -\frac{1}{2q} U$$

$$T_y = \int_0^\infty v \, dz = U \int_0^\infty e^{-qz} \sin qz \, dz = -\frac{1}{2q} U$$

so

$$T_x = T_y = -\sqrt{\frac{v}{2\omega}}\, U$$

so the resultant volume transport is

$$T = (T_x^2 + T_y^2)^{\frac{1}{2}} = U\sqrt{\frac{v}{\omega}}$$

and is at 45° to the right of the surface wind velocity as referred to the upper air.

The induced momentum over unit area of the surface is

$$P = \rho T = U \cdot \sqrt{\frac{\rho\mu}{\omega}} = \frac{1}{\omega} \cdot U \cdot \sqrt{(\rho\mu\omega)} = s/\omega,$$

thus the stress induces a momentum in the lower layer of air, and the stress on the water, being equal and opposite, must induce an equal and opposite momentum on the upper layers of the water. This assertion being based on the conservation laws is not dependent on our assumption about the constancy of v and μ with height. It should be mentioned here that since the stress formula contains $\sqrt{\omega}$, the stress is latitude dependent. In fact for a given geostrophic wind velocity, the stress must vary as $1/\sqrt{(\sin \text{latitude})}$. The mixed layer has a greater height, and du/dz and dv/dz have lesser values as the equator is approached.

It is also worth emphasising that though we saw that

$$S = U \sqrt{(\omega\mu\rho)}$$

this does not necessarily imply that the stress varies linearly with U. It will appear in due time that turbulence must increase with U and so μ also increases with U in a manner to be examined.

THE PATTERN OF MOVEMENT IN THE SEA

There is no need to repeat the mathematical arguments given above. It will be evident that with the appropriate values of ρ and v for the sea, the equations of motion are identical to those for the atmosphere. The resulting pattern of both air and water movement is shown in Fig. 7(a) from the point of view of an observer in the surface and in (b) as a combined figure showing the velocity *changes* induced in each medium by the stress. Figure 7(b) will be useful in emphasising the symmetry and balance of momentum changes in the two media.

Suppose the induced surface drift has a velocity V. From the point of view of an observer in the surface, $-V$ is the velocity of the deep water just as

3. WIND CURRENTS IN DEEP WATER

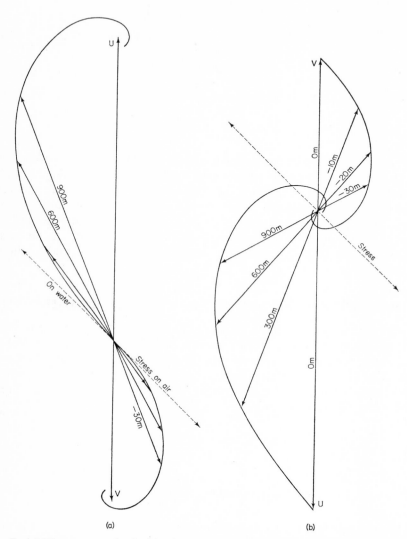

Fig. 7. (a) The pattern of velocities in air and water referred to the surface. (b) The pattern of velocity changes in air and water, relative to the undisturbed velocity in each.

U is the velocity of the upper air. Thus the stress at the surface of the water is

$$S_w = -V\sqrt{(\omega \mu_w \rho_w)}$$

this must be equal and opposite to the stress in the air at that level which we saw to be

$$S_a = U\sqrt{(\omega \mu_a \rho_a)}.$$

Thus

$$\frac{V}{U} = \sqrt{\frac{\mu_a \rho_a}{\mu_w \rho_w}}$$

and this relationship (which does *not* contain ω, and is therefore independent of latitude) determines the surface drift velocity induced by the geostrophic wind. Reference to Fig. 7 will show that the surface drift is in the same direction as the geostrophic wind. It will also be seen that there is a net water flow at 45° to the right of the geostrophic wind, and a net air flow at 45° to the left. Thus all the various momenta balance out, as they must if the original force is derived wholly from the pressure gradient in the atmosphere.

THICKNESS OF THE MOVING LAYERS IN AIR AND WATER

So far we have not discussed explicitly the height in the air or depth in the sea to which interaction motion extends. The formal answer is, of course, to infinity, but a useful yardstick is the height, or depth, at which the motion is rotated by η radians relative to the stress line in the interface. This condition also implies that the velocities relative to the infinity points are $e^{-\pi}$ of U and V. At this level the angle $qz = \pi$, i.e.

$$z = h \text{ (say)} = \sqrt{\frac{2\pi \mu_a}{\rho_a \omega}} \quad \text{(air)}$$

$$-z = \Delta \text{ (say)} = \sqrt{\frac{2\pi \mu_w}{\rho_w \omega}} \quad \text{(water)}.$$

We can thus write the volume transport equations in the convenient form

$$T_x = T_y = -\frac{1}{2\pi} Uh$$

$$T_x = T_y = \frac{1}{2\pi} V\Delta$$

3. WIND CURRENTS IN DEEP WATER

and from obvious considerations of momentum

$$\frac{1}{2\pi} \rho_a U h = \frac{1}{2\pi} \rho_w V \Delta \quad \text{or} \quad \frac{V\Delta}{Uh} = \frac{\rho_a}{\rho_w} \approx \frac{1}{740}.$$

ENERGY CONSIDERATIONS

The rate of energy dissipation per unit volume by the shear against the turbulent forces in air is

$$\mu_a (u_a^2 + v_a^2) = 2\mu_a U^2 q^2 e^{-2qz}.$$

Thus the total energy dissipated per second in the air is

$$W_a = 2\mu_a U^2 q^2 \int_0^\infty e^{-2qz} = \mu_a U^2 q$$

this reduces to

$$W_a = \frac{1}{\sqrt{2}} \sqrt{(\omega \mu_a \rho_a)} \cdot U^2 = \frac{1}{\sqrt{2}} SU$$

as might have been expected.

In the water we have similarly

$$W_w = \frac{1}{\sqrt{2}} \sqrt{(\omega \mu_w \rho_w)} \cdot V^2 = \frac{1}{\sqrt{2}} SV$$

so the ratio

$$\frac{W_w}{W_a} = \frac{V}{U} = \sqrt{\frac{\mu_w \rho_w}{\mu_a \rho_a} \cdot \frac{V^2}{U^2}}$$

The mean rate of energy dissipation per unit volume can be worked out for some finite height and depth. If we take the mean power in the air between 0 and h and in the water between 0 and Δ we have

$$\overline{W}_a = \frac{1}{2\pi} \omega \rho_a U^2 (1 - e^{-2\pi})$$

$$\overline{W}_w = \frac{1}{2\pi} \omega \rho_w V^2 (1 - e^{-2\pi})$$

and putting in the fact that

$$\frac{V}{U} = \sqrt{\frac{\rho_a \mu_a}{\rho_w \mu_w}}$$

we have

$$\frac{\overline{W}_w}{\overline{W}_a} = \frac{V^2 \rho_w}{U^2 \rho_a} = \frac{\mu_a}{\mu_w}.$$

This is as far as it is useful to proceed with deductions from this rather artificial model. It is necessary to look at some empirical results for wind driven currents, and for wind stress, to see what values can, with propriety, be fitted into the model to give some useful guide to real conditions. There is a clear temptation to assume that in the absence of other mixing processes, the balance of power dissipated in the air and water will so influence the degree of turbulence and so the eddy viscosities as to yield more definitive relationships based on the known densities and molecular viscosities of air and water. This however proves to be a complex exercise, and cannot usefully be done with this model. The empirical facts are that the surface water movement is of the order of 2% of the geostrophic wind speed; the stress, as measured from the slope induced on the surface of a shallow sea, is approximately proportional to the square of the wind speed ($S \approx 1 \cdot 86 \times 10^{-3} U^2 \rho_a$); the depth and height of the mixed layers are approximately proportional to the wind speed, though almost certainly less at very low wind speeds.

On this basis it is tempting to introduce a pair of constants k_a and k_w and to write

$$\mu_a = \rho_a k_a^2 U^2$$
$$\mu_w = \rho_w k_w^2 V^2.$$

Here k^2 has the dimensions of time and is to be interpreted as the inverse of a velocity gradient; i.e. as a distance between velocity contours or metres per (metre per second). This gives us for the basic velocity relationship

$$\frac{V}{U} = \sqrt{\frac{\rho_a^2 k_a^2 U^2}{\rho_w^2 k_w^2 V^2}}$$

or

$$\frac{V}{U} = \sqrt{\frac{k_a \rho_a}{k_w \rho_w}}$$

giving us as a practical approximation

$$\frac{k_a}{k_w} = 0 \cdot 25 \quad \text{and} \quad \frac{V}{U} \approx 0 \cdot 0175.$$

For the stress at the surface, as deduced from the air, we have, introducing the empirical results,

$$S = U \sqrt{(\omega \mu_a \rho_a)} = U^2 \rho_a k_a \sqrt{\omega} \approx 1 \cdot 86 \times 10^{-3} \rho_a U^2 \text{ in } 30° \text{ latitude}$$

whence
$$k_a^2 \omega = 3.5 \times 10^{-6}$$
so
$$k_a^2 = 0.1 \qquad k_a = 0.32$$
and
$$k_w^2 = 1.6 \qquad k_w = 1.3.$$

For the mixing height in the atmosphere we have

$$h = \sqrt{\frac{2\pi\mu_a}{\omega\rho_a}} = 0.8\,U\sqrt{\frac{1}{\omega}} \approx 130\,U \text{ in } 30° \text{ latitude.}$$

For the mixing depth in the water we have

$$\Delta = \sqrt{\frac{2\pi\mu_w}{\omega\rho_w}} = 3.2V\sqrt{\frac{1}{\omega}} \approx 530V \approx 9\,U \text{ in } 30° \text{ latitude.}$$

EMPIRICAL CONCLUSIONS REGARDING TRANSPORT AND ENERGY

We can use these results to put figures of a tentative sort into some of the more important formulae previously deduced. First, we have the transport equations:

$$T_x = T_y = -\frac{1}{2\pi}Uh \qquad \text{(air)}$$

and

$$T_x = T_y = \frac{1}{2\pi}V\Delta \qquad \text{(water).}$$

These reduce to

$$T_x = T_y = 0.13\,U^2\sqrt{\frac{1}{\omega}} \approx 21U^2 \qquad \text{in latitude } 30°$$

$$T_x = T_y = 0.51V^2\sqrt{\frac{1}{\omega}} \approx 85V^2 \approx 0.03U^2 \text{ in latitude } 30°.$$

Thus the transport of air and water is proportional to the square of the geostrophic wind, and a doubling of the wind speed results in doubling both the speed and the thickness of the moving layers. The empirical results agree in general with this conclusion, except that at low wind speeds when the sea surface is smooth, the stress and depth of mixing is significantly less than indicated by these simple formulae (see e.g. Sverdrup *et al.*, 1942).

The total rate of energy dissipation in a vertical column of 1 square metre area, works out on this model as

$$W_a = \frac{1}{\sqrt{2}} \sqrt{(\omega \mu_a \rho_a)}\, U^2 \approx 0.28 U^3 \sqrt{\omega}$$

$$\approx 1.7 \times 10^{-3}\, U^3 \text{ watts in } 30° \text{ latitude}$$

$$W_w = \frac{1}{\sqrt{2}} \sqrt{(\omega \mu_w \rho_w)}\, V^2 \approx 5.5 V^3$$

$$\approx 3 \times 10^{-5}\, U^3 \text{ watts in } 30° \text{ latitude.}$$

Thus the rate of dissipation of energy by motion against the viscous forces varies with the cube of the velocities concerned.

LIMITATIONS OF THE EKMAN MODEL

The preceding pages have been occupied by discussion of the very famous theory developed by V. W. Ekman in the early years of the century. The ideas give a useful introduction to the problem, and an elegant concept of the way the wind and water stresses must balance each other. The assumption that the viscosity is independent of height and depth is clearly not an accurate one, and it is right also to be highly suspicious of the variations of the various quantities with latitude, at least in the way they have been developed here. It is instructive to look again at some of our formal conclusions to see which are likely to have absolute validity irrespective of the assumptions made heretofore.

We may express the situation as follows: suppose an infinite ocean of large depth, and at rest, as is also the air above it. If we now impose a pressure gradient which we shall denote as a force F per unit volume, this generates a momentum F/ω per unit volume on air and sea alike. Since the water density exceeds the air density by a factor of about 750, the directly-induced air velocity exceeds the directly-induced water velocity by this factor. In fact, we may ignore the directly-induced water velocity for most purposes.

At the interface there is a stress generated due to the difference in velocity. Since the momentum rule is absolute, for steady state conditions, the stress can make no difference to the total momentum generated by F. Instead, the stress acts by generating equal and opposite supplementary momenta in the air and water.

It will be recalled that the velocity spiral in the air (or in the water) is such that in some small increment δ of height the vector of induced velocity decreases as $e^{-q\delta}$ and the direction of the vector changes by an angle $q\delta$

radians. As a result of this relationship the spiral was a 45° one, and an observer moving with the wind at any level experienced a stress at 45° to the apparent wind at large height. It will be evident that even if q varies with height, this relationship will remain true. The effect of change in q is to vary the height increment δ associated with this particular change in velocity and angle. If q varies with height, that is to say if μ varies with height, the Ekman spiral will continue to apply, but the vertical spacings associated with steps in direction will no longer be equal.

Thus the stress *will* be at 45° to the geostrophic wind. The empirical stress values are unaffected by any assumption about constancy of μ, and the momentum transports, being dependent on the stress, will also be correct. However, our figure for mixing height and depth and our conclusions generally about the vertical distribution of all properties, become convenient fictions approximating only in general terms to the true situation.

ALTERNATIVE VIEW OF THE EKMAN SPIRAL

There may seem to some readers to be a slickness about this description of the effect of wind on water, which seems to lack self-evidency. An alternative view is to recognise that momentum is imparted to the sea surface, by the stress of the wind. This momentum diffuses downward into the sea (at a rate proportional to μ). Near surface the effect of the most recent stress is dominant. At great depth the momentum has been derived from stress spread over a long period of several pendulum days, and the momentum there is derived from all possible true directions and so cancels.

At a depth so chosen that the momentum is principally derived from stress half a pendulum day earlier, the momentum is obviously opposite to that currently being imparted at the surface.

Thus we see that the effect of the earth's rotation is to confine wind-driven currents to a surface layer, whose depth is dependent on the viscosity of the water and indirectly on the intensity of the wind stress. A cyclonic wind pattern may be expected to produce an outward transport of the upper water and induce upwelling in the centre, if stability is not too great.

Stress obviously depends on the roughness of the sea surface and, as waves develop the turbulence will increase, giving an increase in the depth of the moving mixed layers.

BIBLIOGRAPHY

Batchelor, G. K. (1967). "An Introduction to Fluid Mechanics", 615 pp. Cambridge University Press.

Chapter 4: Qualitative Physical Oceanography

WATER MOVEMENTS IN A BOUNDED SEA

Our consideration of density and wind-driven currents has been directed to highly idealized situations. The most important considerations which have been omitted are concerned with the finite nature of the earth, with the fact that the oceans and seas are in various degrees interspersed with land, which limit the kind of movement possible, and the scale of meteorological patterns each of which is separately trying to impose its characteristic motion on the upper waters.

In the previous chapter we quoted an empirical formula for the wind stress at the sea surface

$$s \approx 1 \cdot 86 \times 10^{-3} U^2 \rho_a$$

giving for a geostrophic wind speed of 10 m sec^{-1} in middle latitudes

$$s = 0 \cdot 23 \, \text{N m}^{-2}.$$

Suppose the depth of the sea were d m then the vertical gravitational force on the column under 1 m^2 of surface would be $dg\rho_w$: i.e. 9810d N and the slope which would just balance the wind stress would be 0·23/9810d or for $d = 100$ m would be $0 \cdot 23 \times 10^{-6}$ or, for clearer visualization, 23 cm per 1000 km.

To reach this slope, water has to flow in the general direction of the stress. We may consider a slice in this direction supposed to be one meter in width (Fig. 1). If β is the slope, the volume to be transported across any section

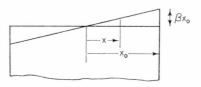

Fig. 1. Slope on a sea surface generated by a wind stress from left to right.

at x will be $v = \frac{1}{2}\beta(x_0^2 - x^2)$ where x_0 is half the length of the slice. Applying this to our example of a sea 1000 km in width we obtain Fig. 2. How long will it take for the slope to come into equilibrium with the stress? This clearly (in this elementary argument) does not depend on the depth to which the movement extends. Suppose this depth is z_0 then since the maximum volume transfer (at the centre of the slice) is 30,000 m³ the horizontal motion involved is $30,000/z_0$. The mass moved under 1 m² is $z_0 \rho_w$ and the acceleration due to the wind is $s/z_0\rho_w$ m sec⁻² and so the distance moved in t sec is

$$\frac{1}{2}\left(\frac{s}{z_0\rho_w}\right)t^2$$

and the time to equilibrium is given by

$$\frac{1}{2}\left(\frac{s}{z_0\rho_w}\right)t^2 = 30,000/z_0$$

whence $t \approx 4\frac{1}{2}$ hr.

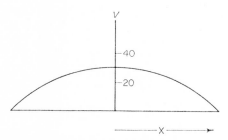

Fig. 2. Volume of water to be transported across a 1 m wide section to reach the uniform slope of Fig. 1, in thousands of cubic metres, for a sea width of 1,000 km, a depth of 100 m, and a wind speed of 10 m sec⁻¹.

Thus, in this instance, net transport due to the wind stress will cease after a time of this order, and any continuing upper water motion must be compensated by a return flow at some lower level, or at least by some return path. Reference back to the argument will show that this "equilibrium time" at the sea centre is proportional to the width of the sea and inversely proportional to its depth.

Obviously the "equilibrium time" becomes less as we approach the weather or lee shore of our sea, being zero at the coastlines. To emphasise this quite elementary point it is interesting to consider what would happen if a uniform transport of the upper z_0 metres were imposed by wind stress on the same sea hitherto studied. After 1 hr the sea levels in our slice would have become as shown in Fig. 3.

The volume transported across each cross section would be

$$v = \frac{1}{2}\left(\frac{s}{\rho_w}\right)t^2 \quad \text{or about } 1600\,\text{m}^3.$$

Since the transport is the same at all points, no slope is generated in the main body of the water. At the weather shore the upper z_0 metres is moved bodily offshore leaving a step in the water surface, while at the lee shore the final infinitesimal slice is raised to an infinite height. There are two reasons for including this absurd example, first to drive home the obvious fact that net transport in the near shore region is always parallel to the coast, the possibility of a net transport perpendicular to the coast being limited to durations of minutes. The second point is that when a wind sets in across the sea, it tends to set up initially a sinuous distortion of levels (see Fig. 3b) and so compensation currents set in first in the coastal region. Only if the wind continues for a period of hours is a complete transport cell set up as in Fig. 3(c).

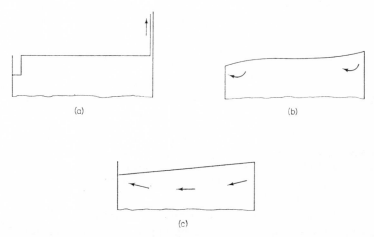

Fig. 3. Illustrating effect of a left to right wind stress on water levels.

EFFECT OF A THERMOCLINE ON CIRCULATION

We saw that the influence of an abrupt thermocline was to decouple the lower water from the wind-driven upper layer so that compensation currents can be expected within the upper layer only (Fig. 4a). In most actual sea conditions however the thermocline is limited to the undisturbed central area, being characteristically broken up by tidal currents in the coastal regions. Thus the situation is by no means simple and the effect of wind driven transport is to combine with other effects in progressively destroying the discontinuity which, formed during the early summer, is continually being "nibbled away" at the edges (Fig. 4b).

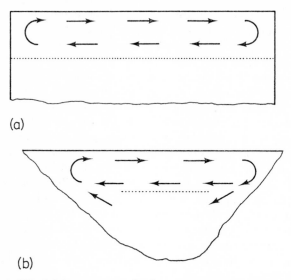

Fig. 4. Wind-driven currents in the prescence of a thermocline.

Since this is a practical chapter, gathering together ideas, it may be permissible to complete this statement by showing, in Fig. 5, another process which is here involved. The coastal water of intermediate density naturally tends to flow between the warm and cold layers of the stratified central region. That it does not do so immediately is to be attributed to the effects of the earth's rotation, which tends to inhibit such processes by turning the

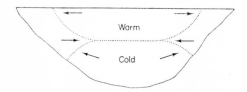

Fig. 5. Self-destruction of a thermocline by density currents.

motion to the right (in the northern hemisphere) and converting it into motion along the density interfaces. If there were no frictional or viscosity forces, it could in principle maintain such systems indefinitely.

Another point worthy of note is that the equilibrium slope of the sea surface, for any given wind stress, is inversely proportional to the depth. Thus, where there is along the line of the wind driven upper water movement a deep and a shallow area, there will be a tendency for the water "heaped up" over the shallows to return to windward via the deeper region. This may or may not be sufficiently extreme to result in an actual surface flow against the wind in the deep region. Whether or not this occurs will depend on the scale of the phenomenon, and on the relationship between the horizontal and vertical coefficients of eddy viscosity. Figure 6 indicates how such a situation can arise when motion induced by the wind is inhibited by a coastline obstructing the wind-driven flow. Clearly, the same situation would arise if the wind were reversed so as to be blowing out of the bay. A lee shore, or a weather shore are equally inhibiting of net transport. Suppose however that instead of a bay, we have an open passage, as in Fig. 7. It might appear that net flow could occur in an uninhibited way. But of course this is not necessarily true either, for the water levels at A and B will be set by the wind situations elsewhere, thus there may or may not be a return flow in the direction from A to B. Indeed what we can say with certainty is that if the situation is maintained for any length of time, there *must* be return flow from A to B, even though this may be by a very devious route of many thousands of miles.

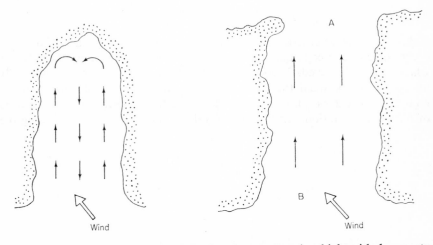

Fig. 6. (Left) Illustrating the tendency of surface currents in a bight with deep centre and shallow sides. **Fig. 7.** (Right) Wind current uninhibited by local land. (See text.)

What these qualitative arguments suggest is the need to consider wind systems on the largest feasible scale before reaching even tentative conclusions about resulting water movement, and the need to check any conclusions by a sufficient mass of direct observations.

ESTUARINE TYPES OF WATER MOVEMENT

Apart from wind, the most efficient propelling forces arise from changes of density. On the medium scale (over hundreds of km as opposed to thousands) such changes of density arise mainly as a result of dilution of the sea by rivers. We do not concern ourselves here with genuine river situations where measurable slopes can occur on the surface even if the density is uniform. Such situations are the province of the hydraulic engineer. Rather let us consider the situation in a large inlet such as might fill from the associated rivers in a few days. If we suppose this to have happened, and the inlet to be entirely fresh water, abutting on a saline sea, then the development of such a situation to the observed type is easy to comprehend. The density in the inlet will be about 1000, and in the sea about 1025 kg m^{-3}. Thus, as described in Chapter 1, the pressure cannot be the same at all depths in each water mass. At a depth $z = 10$ m below the surface the pressure difference will be $\rho'gz$ where ρ' is the density difference. This amounts to $10 \times 9\cdot8 \times 25 = 2{,}400$ N m^{-2} or roughly a quarter of a ton per square metre. Evidently therefore the system is not in equilibrium and the salt water will force its way under the fresh, displacing fresh water which will bulge out as shown by the dotted line.

The system would develop into the observed type as shown in Fig. 8(b) in which the following processes take place:

The surface is slightly higher at the head of the inlet than in the sea, on account of the differing densities. The "brackish" upper layer moves seaward, eventually to be dispersed. As it moves seaward it gains in salinity by turbulent mixing with the intruding saline water beneath it. The volume flow increases as the stream passes seaward e.g. suppose the fresh water flow were 10 m^3 sec^{-1}, then at a point in the stream where the fresh water is in the ratio of 1 part to 10 parts of sea water, the outward volume flow must 110 m^3 sec^{-1}. We know that this must be so if the situation is in equilibrium as all the fresh water must be removed to seaward. Finally the removal of sea water by the outflow must induce an equal inward flow in the lower water. For instance at the section where the outflow is 110 m^3 sec^{-1} the inflow will be 100 m^3 sec^{-1}. This somewhat simplified description brings out the fact that the continuing addition of quite a small volume of fresh water can induce sympathetic motion of very much larger volumes of sea water.

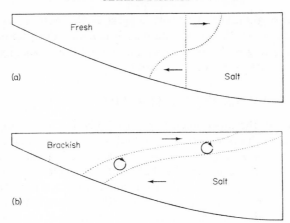

Fig. 8. Development of estuarine circulation.

As a result of Coriolis' force the interface between the out and inflowing masses is inclined but this inclination is in general so slight that it is not very significant until the inlet has widened to around 3 km. Beyond this sort of width, the characteristic effect is a segregation of the outflowing diluted water to the upper right of the channel, as in Fig. 9. Even a large water body like the North Sea can be roughly described in terms of this estuarine form. A west–east section of salinity would look quite closely like Fig. 9. On a

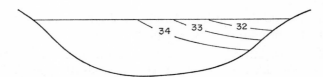

Fig. 9. Salinity distribution typical of an outer estuary in the Northern hemisphere. Looking seaward, the low salinity water flows out in the upper layers along the right-hand shore. The remainder of the section will show a net inflow of more saline water from the sea.

larger scale the circulation of the Northern Ocean can be similarly considered, most of the fresh water coming from the melting of sea ice and of glacier ice from the Greenland cap.

EFFECT OF ICE FORMATION

Just as the melting of sea ice gives rise to dilution of the surface water, so also is there an inverse process when freezing takes place. The salt content of sea ice is less than that of the water from which it derives. It is not zero because salt is typically trapped in pockets within it, but the freezing process nevertheless raises the salinity of the immediately adjacent sea water, and raises its density to such an extent as to cause it to sink. Thus much of the cold bottom water of the oceans derives from the formation of ice, while many of the really cold surface currents derive from its melting.

The conditions for ice formation on the sea surface are not easy to define. Unlike fresh water, which has its maximum density at around 4°C, sea water of salinity greater than 15‰ goes on increasing in density as the temperature decreases, right down to freezing point. As a warm fresh water lake is chilled by adverse radiation balance, by evaporation, and by contact with cold air, convection currents set in and the temperature of the whole water column is reduced to 4°C. If the chilling continues beyond this point, convection currents cease, and the surface water cools off to the freezing point, leaving the deep water mass at 4°C.

By analogy it would appear that freezing of the sea would not take place until the whole water mass is cooled to freezing point (around 1·9°C for water at 35‰). However we know that this is too stringent a condition. Examination of the figures in the appendix shows that the change in density of sea water between 0° and 5°C is just about equal to that caused by a change of salinity from 35‰ to 34·5‰. Thus if the surface water is freshened to this extent by rivers, or by local precipitation, then freezing can take place at the surface.

Ice formed on the sea surface does not normally exceed about 4 m in thickness, except where floes are crushed together. Substantial icebergs, which may be many hundreds of metres in thickness are derived entirely from the great ice caps of Antarctica and Greenland. Such ice-loaded currents sweeping into temperate regions owe much of their effectiveness to the latent heat of the ice they carry, and modify the climate of the regions they bathe.

BIBLIOGRAPHY

von Arx, W. S. (1962). "An Introduction to Physical Geography", 422 pp. Addison-Wesley, Reading, Mass.

Sverdrup, H. U., Johnston, M. W. and Fleming, R. H. (1942). "The Oceans, their Physics, Chemistry and General Biology", 1087 pp. Prentice Hall, Englewood Cliffs.

Chapter 5: Waves in Deep and Shallow Water

WAVES IN DEEP WATER

Superficial observation shows that surface waves are produced when wind blows over the sea. While the wind continues the wavelength and height of the waves increase with the distance travelled, up to some limit dependent on wind velocity. When the wind ceases, the waves continue with apparently unchanged wavelength but gradually decreasing height. The velocity of waves increases with the wavelength.

The sea surface is of complex pattern, a spectrum of wavelengths being normally present, and any slight differences in direction of component waves has the result of breaking up the uniform crest lines by interference, yielding typical "short crested" waves.

To understand the processes involved, we have to idealize the situation, and we shall first consider a wave train of constant amplitude and wavelength. How do we account for the form? How does the velocity relate to the wavelength? How deep does the disturbance extend? Is the group velocity different from the phase velocity?

It is useful to start with the assumption that the surface form of a self-maintaining wave is cycloidal, as shown in Fig. 1. Each surface particle rotates in a circle of radius A (thus the amplitude A is half the wave height). The phase of the rotation changes with distance along the wave train. Thus, to be precise, if we take the positive direction of x to be the direction of the wavemotion and we define any particle by the x-coordinate it would have in the absence of wavemotion, then we see that in the presence of the wave it finds itself at the head of a radius vector of length A, making an angle θ with the forward horizontal direction, where

$$\theta = -\omega t + 2\pi x/\lambda.$$

Fig. 1. Movement of surface particles in a progressive wave.

5. WAVES

The justification for assuming a cycloidal form must lie in the degree to which such a form fits the observed facts. It should be pointed out that the postulated motion can be regarded as the sum of two simple harmonic motions, one vertical with amplitude A, and one horizontal, also of amplitude A and differing in phase by $\pi/2$. Thus we are justified in combining wavetrains and studying interference effects by the methods used for simple harmonic motion.

WAVELENGTH AND PHASE VELOCITY

The fundamental idea is that the water surface sets itself perpendicular to the resultant force. This resultant force on a surface particle is the combination of the gravitational force mg, and the centrifugal force $m\omega^2$ due to the rotary motion ($\omega = 2\pi f$ where f is the wave frequency). It is most convenient to describe the point of interest on the wave surface by the phase angle θ corresponding to it. We can determine the deflection from the vertical of the resultant force R by reference to Fig. 2(a). It is evident that

$$\tan \psi = \omega^2 A \cos \theta / (g - \omega^2 A \sin \theta)$$
$$= A \cos \theta / (g/\omega^2 - A \sin \theta).$$

We may next consider the wave slope at the same point in the surface. Referring to Fig. 2(b) we see that the vertical distance from P to Q is

$$A \, d(\sin \theta) = A \cos \theta \, d\theta.$$

The horizontal distance from P to Q is

$$dx + A \, d(\cos \theta) = dx - A \sin \theta \, d\theta.$$

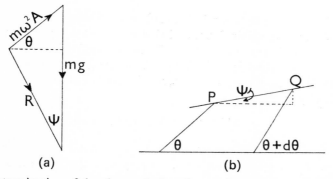

Fig. 2. Determination of the phase velocity of waves in deep water. (a) Inclination of resultant force. (b) Slope of surface.

But
$$\theta = -\omega t + 2\pi x/\lambda$$

so for t constant we have

$$d\theta = 2\pi \, dx/\lambda \quad \text{or} \quad dx = \lambda \, d\theta/2\pi.$$

Thus
$$\tan \psi = A \cos \theta d\theta/(dx - A \sin \theta d\theta)$$
$$= A \cos \theta/(\lambda/2\pi - A \sin \theta).$$

We now have to test the possibility that these two results can be equated for any value of the phase angle θ i.e. we must have

$$A \cos \theta/(\lambda/2\pi - A \sin \theta) = A \cos \theta/(g/\omega^2 - A \sin \theta)$$

and these are evidently equal for all values of θ if

$$\lambda/2\pi = g/\omega^2. \tag{1}$$

The phase velocity of any wavemotion is given by

$$c = f\lambda = \omega\lambda/2\pi$$

that is
$$\omega^2 = (2\pi c/\lambda)^2.$$

Substituting this value in equation (1) we obtain

$$\lambda/2\pi = g \, \lambda^2/4\pi^2 c^2$$

that is
$$c^2 = g \, \lambda/2\pi \quad \text{or} \quad c = \sqrt{(g\lambda/2\pi)}.$$

Thus we have shown (by considering surface conditions only) that a cycloidal surface wave is a possible form, and that the phase velocity should have a certain relationship to the wavelength. This calculated value is in good accord with observation, which is not in fact particularly easy to carry out accurately. For example we may consider the case of waves of 100 m wavelength. The phase velocity is

$$c = \sqrt{(100 \times 9\cdot 81/2\pi)} = 12\cdot 5 \text{ m sec}^{-1}.*$$

While there are other limitations to be considered on the amplitude of sea waves of any given wavelength, we may note that there is an absolute

* See note added in proof, p. 54.

5. WAVES

limit given by the relation $A\omega^2 = g$. If a wave should exceed this amplitude, the crest would become separated from the main body, and the wave could not continue as a progressive wave. This outside limit is found by substituting for ω^2 its value $2\pi g/\lambda$ from the equations above. Thus

$$A \cdot 2\pi g/\lambda = g \quad \text{or} \quad A_{max} = \lambda/2\pi$$

If the appropriate cycloidal surface is drawn up for this limiting case, it will be found that the wave crests become peaks: i.e. there is a discontinuity in the wave slope at each crest. Although progressive waves of this amplitude and steepness are not found at sea, it proves convenient to consider such limiting amplitudes in the study of the disturbance beneath the surface, which is naturally associated with surface waves.

SUBSURFACE WAVEMOTION

The argument relating the slope of the wave surface with the deflection of the resultant force applies to any isobaric surface equally as it does to the surface which is merely one particular isobaric surface. Hence we may examine the possibility of cycloidal waves on the subsurface isobaric surfaces, and consider how the amplitude of these would vary with depth. It is evident that the amplitude at a subsurface isobar must be less than at the surface. This is because the resultant force on a particle at the wavecrest is $m(g - \omega^2 A)$ hence

$$dp = (g - \omega^2 A) \rho \, dz':$$

that is the rate of pressure increase with depth is *less* than for undisturbed water. Under a trough

$$dp = (g + \omega^2 A) \rho \, dz''$$

that is the rate of pressure increase with depth is *greater* than for undisturbed water. Thus to reach the pressure corresponding to one metre depth in undisturbed water, we have to go down from a crest more than one metre, and from a trough less than one metre. So the wave amplitude at the one metre isobar must be less than at the surface. If we consider more fully the situation under a crest, we have

$$dz' = \frac{1}{\rho (g - \omega^2 A)} dp$$

where z' is the depth from the crest to the p isobar. Now in undisturbed water we have

$$dz = \frac{1}{\rho g} dp.$$

The difference between these values is the decrease in amplitude corresponding to the pressure change dp, that is

$$dA = -(dz' - dz) = -\frac{1}{\rho} dp \left(\frac{1}{g - \omega^2 A} - \frac{1}{g} \right)$$

hence

$$(g/\omega^2 A - 1) dA = -\frac{1}{g\rho} dp$$

so

$$g/\omega^2 \log A/A_0 - (A - A_0) = -p/g\rho$$

since

$$A = A_0 \quad \text{for} \quad p = 0$$

that is

$$g/\omega^2 \log A/A_0 - (A - A_0) = -z$$

or since $g/\omega^2 = \lambda/2\pi$

$$\log A/A_0 = -2\pi/\lambda \, (z + A_0 - A).$$

Now we have been measuring from surface crest to isobar crest. If we choose to measure instead from the centre of rotation of surface particles to the centre of rotation of particles in the p isobaric surface, this becomes simply

$$\log A/A_0 = -2\pi z/\lambda.$$

If we assume that this surface wave is a maximum wave, we can put $A_0 = \lambda/2\pi$ and derive a general expression for the penetration of wave disturbance into the sea, that is

$$A/A_0 = \exp\left(-\frac{2\pi z}{\lambda}\right)$$

or

$$A = \lambda/2\pi \exp\left(-\frac{2\pi z}{\lambda}\right).$$

A similar argument based on the depth of isobars beneath a trough leads to precisely the same conclusion. We can now construct Fig. 3, showing the way in which wave amplitude decreases with depth. This is perfectly general for we can take as surface a level corresponding to any selected wave amplitude, and measure down in wavelengths to find the depth at which the amplitude has decreased to any particular value. The easy way to prepare such a diagram is to assume a wavelength of 2π m: that is 6·28 m, giving an A_{max} of one metre. It is almost impossible to grasp this subject clearly without constructing both this Figure, and the following Fig. 4.

In this figure are shown most of our conclusions about simple gravity waves. The instantaneous shapes of the various isobaric surfaces are of

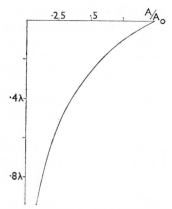

Fig. 3. Decrease with depth of the amplitude of a deep-water wave.

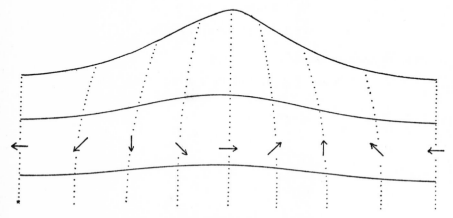

Fig. 4. Isobars and isophase lines in a deep-water wave.

course cycloids. By erecting verticals at $x = 0$, $\lambda/8$, $\lambda/4$, $3\lambda/8$ etc. and offsetting an amount to allow for the length of the amplitude vector when it is horizontal, or inclined at 45° we can sketch in the isophase lines, and hence the directions of particle motion. The basic rule will be apparent, that *velocity* of motion is identical for any isobar, and that *direction* of motion is identical for any isophase line. Just as in Fig. 3, it should be noted that any isobaric surface can be assumed to be the sea surface and the layers below will show the appropriate subsurface situation.

WAVES IN SHALLOW WATER

Here we shall consider the extreme situation where the depth z is very much less than a wave length. For simplicity we assume the wave amplitude h to be also small. We shall not go into any great detail but will be content to show that in such circumstances the phase velocity follows an entirely different law than for the case of deep water waves, being dependent only upon the depth, and being independent of the wavelength provided only the above conditions remain satisfied.

In dealing with this problem one of the fundamental theorems of hydrodynamics, which is due to Bernoulli, is needed. For streamline flow we can consider the conditions applying *to tubes of flow*, i.e. tubes supposed parallel to the streamlines (Fig. 5). We make the assumption that there are no frictional losses, and so the work done on the water in the tube will be equal to the energy gained by the water. This energy consists of course of the kinetic energy and the potential energy.

Let a small volume Δ flow through the tube in the direction shown. Then we have

$$\text{Work done on system} = p_1 \Delta$$

$$\text{Work done by system} = p_2 \Delta$$

so net work done on system $= p_1\Delta - p_2\Delta$

Mass of Δ is $\rho\Delta$ where ρ is the density

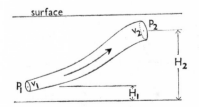

Fig. 5. Illustrating Bernoulli's theorem.

so

$$\text{gain in kinetic energy} = \tfrac{1}{2}\rho\varDelta v_2{}^2 - \tfrac{1}{2}\rho\varDelta v_1{}^2$$

and

$$\text{gain in potential energy} = g\rho\varDelta H_2 - g\rho\varDelta H_1$$

Thus since

work done on system = gain in energy of system

that is $p_1\varDelta - p_2\varDelta = \tfrac{1}{2}\rho\varDelta v_2{}^2 - \tfrac{1}{2}\rho\varDelta v_1{}^2 + g\rho\varDelta H_2 - g\rho\varDelta H_1$

and by sorting out the terms and cancelling we obtain

$$\frac{p_1}{\rho} + \tfrac{1}{2}v_1{}^2 + gH_1 = \frac{p_2}{\rho} + \tfrac{1}{2}v_2{}^2 + gH_2$$

or as a more formal statement—along any tube of flow

$$\frac{p}{\rho} + \tfrac{1}{2}v^2 + gH = \text{constant.}$$

Let us now consider the application of this general theorem to our wavemotion (Fig. 6). For simplicity of treatment we take the standard method of adopting a frame of reference which moves with the wave profile at the phase velocity c. Thus, referred to this frame, the water flows to the left under the wave profile at a mean velocity c. However it passes through a shallower cross section under a trough than it does under a crest and so its velocity is higher under the trough. Let us call the velocity under the trough $c + v$ and under the crest $c - v$. (This means that in an earth-referenced frame v is the peak value of the oscillatory particle velocity).

From the diagram it can at once be seen that

$$\frac{c+v}{c-v} = \frac{z+h}{z-h}$$

or more simply that

$$\frac{c}{v} = \frac{z}{h}. \tag{1}$$

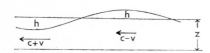

Fig. 6. Wavemotion in shallow water.

Applying Bernoulli's theorem to a tube of flow in the surface layer from crest to trough we have

$$\frac{p}{\rho} + \tfrac{1}{2}(c+v)^2 + g(z-h) = \frac{p}{\rho} + \tfrac{1}{2}(c-v)^2 + g(z+h)$$

and since p is atmospheric pressure and so constant,

$$\tfrac{1}{2}\{(c+v)^2 - (c-v)^2\} = g\{(z+h) - (z-h)\}$$

or

$$\tfrac{1}{2}(4cv) = 2gh$$

or

$$cv = gh \qquad (2)$$

now from equation (1)

$$h = \frac{vz}{c}$$

so (2) becomes

$$cv = g\frac{vz}{c}$$

or

$$c^2 = gz \quad \text{or} \quad c = \sqrt{(gz)}$$

Thus c depends only upon the depth.

As a check we may consider a tube of force directed along the sea bed, from a point under the crest where the pressure is p_c to a point under the trough where it is p_t. In this case the vertical height is zero at both ends of the tube, so the theorem requires

$$\frac{p_t}{\rho} + \tfrac{1}{2}(c+v)^2 = \frac{p_c}{\rho} + \tfrac{1}{2}(c-v)^2. \qquad (3)$$

But $\qquad p_t = g\rho(z-h) \qquad$ and $\qquad p_c = g\rho(z+h)$

so the equation becomes

$$g(z-h) + \tfrac{1}{2}(c+v)^2 = g(z+h) + \tfrac{1}{2}(c-v)^2$$

which is the same as equation (1) and therefore confirms the previous conclusion.

VALIDITY

The simplicity of the treatment depends on our selection of z and h small compared with λ. This means that the treatment is suitable for consideration of long shallow waves, such as coastal tides with wavelengths of the order of hundreds of miles. In such a case we can see that the fundamental difference from deep water waves is that the orbital motion of the water particles has been replaced by a to-and-fro almost purely horizontal motion. We can see that the particle motion is in the same directoin as the wave under the crest but is opposed to it under the trough and, in this respect, such shallow water waves are conformable to deep water waves.

The particle velocities can be computed easily if in equation (2) we substitute for c instead of v. Thus

$$cv = gh \quad \text{and} \quad c = \frac{vz}{h}.$$

Hence the peak particle velocity is given by

$$v = h\sqrt{\frac{g}{z}}.$$

It should be noted that in this treatment no special assumption was made about the wave profile. It will be easily seen that provided the slope is not too steep, the same argument can be applied to the velocity of transmission of a surge or of an isolated wave.

TRANSITION EFFECTS

No precise treatment will be attempted of the complex situation which arises when deep water waves run into shallow water. However, an interesting point may be mentioned, which is easily confirmed by observation. Since in the shallow water situation the phase velocity of a wave train is independent of the wavelength, the group velocity is no longer different from the phase velocity. Thus it was mentioned that a wave in deep water has no continuing individuality. If we fix our eyes on a crest and follow it, soon it will lose its identity and become a trough. However, if we watch heavy seas running into shallow water, this phenomenon no longer holds. and starting from a a certain distance an individual wave *can* be followed, retaining its identity as a relatively high or low wave, until finally it turns over and releases its energy in a turmoil of foam.

Also, since the velocity in shallow water is controlled by the depth, refraction can occur as this depth varies, and beautiful patterns can be observed, perhaps best exemplified by the manner an onshore swell enters a bay from seaward. If the bay is a typically sandy one, deepest in the centre, the successive wave crests of an even swell proceed more rapidly in the deep water and the wave front becomes convex in plan, as if to fit its contour to the form of the receiving shore.

BIBLIOGRAPHY

Braddick, H. S. J. (1965). "Vibrations, Waves and Diffraction", 217 pp. McGraw-Hill, New York.
Lamb, H. (1924). "Hydrodynamics", 687 pp. Cambridge University Press.
Russell, R. C. H. and Macmillan, D. H. (1952). "Waves and Tides", 348 pp. Hutchison's Scientific and Technical Publishers, London.

NOTE ADDED IN PROOF

GROUP VELOCITY

Because c depends on λ, the group velocity V_g is not the same as the phase velocity. There is a well known relationship (e.g. Braddick, 1965)

$$V_g = c - \lambda \frac{dc}{d\lambda}$$

and if we substitute our value for c we find that

$$V_g = \tfrac{1}{2}\sqrt{(g\lambda/2\pi)} = \tfrac{1}{2}c$$

that is, the group velocity is exactly one half of the phase velocity.

The group velocity is the rate of transport of energy, and the rate we would use to predict the arrival of waves from a remote storm centre. If we look at a simple wave group, we see the individual waves develop at rear and work their way forward to disappear at the leading edge. The group is an entity which can be followed for long distances, whilst the individual wave is evanescent, its lifetime depending on the group length or, which amounts to the same thing, to the homogeneity of the wavelengths present.

Chapter 6: The Tides

Everyone has some concept of the tide and those who live by the sea acquire detailed knowledge in relation to their home ground. It is therefore tempting to break from the dogmatic brevity which has characterized the approach of this book, and to treat this subject in a more inductive manner. At most places, the characteristic feature is of two high waters and two low waters in 25 hr approximately. Thus high water on one day at 10·00 in the morning will be followed by evening high water at around 22·30, and the next morning's tide will be at about 11·00.

Secondly, one is aware that small tidal ranges (so-called neap tides) are followed after an interval of seven days by large tidal ranges (spring tides) in a fairly smooth progression. Generally the spring tidal range is about twice the neap range. After a few year's observation it will become noticeable that the spring range itself varies by about 30%, and this makes itself particularly obvious by the exposure of shoals or banks to an exceptional degree. The variation in the neap range will not strike the imagination so forcibly unless actual measurements are made.

Depending on where we are, we may or may not notice that successive tides (morning and evening) can be significantly different in height. This phenomenon is by no means striking in some areas, though important in others.

Finally, an observer limited to one area will probably notice, if he tries to be at all careful in his observations, that the "one hour later each day" rule is far from exact. In fact the time of high water will advance only about 30 min per day during the period of spring tides, and will then race forward by about 70 min per day during neap tides. As a result of this effect, if we take as reference the times of high water at spring and neap tides, and interpolate between these to find the time of other high waters, we shall find the midway tide after springs preceding our predictions by about an hour, and the midway tide after neaps lagging on our prediction, again by about an hour.

If observation is not confined to one place, but is extended to many (which is of course most easily done with the aid of a set of tide tables) it will be found that the variation in tidal range is considerable. Around the

British mainland alone we find places such as Lowestoft, where the spring range varies between 2·0 and 2·7 m, and at the other extreme Avonmouth, the port of Bristol, where the spring range varies between 10·9 and 14·5 m. If we consider likewise the minimum predicted rise or fall for these places at neap tides, we find for Lowestoft 0·5 m, and for Avonmouth 5·4 m. Thus the variability of tidal range from place to place and time to time is perhaps greater than is generally appreciated.

Other points might be noted. For instance, when we find exceptionally large spring tides, the preceding and subsequent spring tides will generally be small. In fact we find periods when big spring tides alternate with small spring tides, followed by periods of more or less uniform ordinary spring tides. An obvious point which perhaps might have been emphasised sooner is that in any particular location there will be a relationship between the time of high water and the range of the tide. Thus at some particular place we might find that afternoon high tides were always neaps. This perfectly definite relationship can have ecological and social significance. For instance, where lowest low waters occur at noon, the exposed shore is subject to more heat and drying than if the lowest low waters occur at 06.00 and 18.00. Similarly, beach activities of the human species are concentrated in the afternoon, and it is on the whole an advantage for a resort if the lowest low tides do not occur then.

THEORY

These qualitative remarks may be useful to those whose interest lies not in the tides themselves, but in the influence which the tides may have on other investigations. This of course includes the greater number of those of us who have to do experimental work at sea. Pursuing the same vein of thought, we shall try to make clear the elementary background to tidal phenomena without attempting any completeness of treatment. The obvious reference for UK readers seeking detail is the Admiralty Manual of Tides (Doodson and Warburg, 1941), and the fundamental paper by A. T. Doodson (1921).

A particle at the centre of the earth is subject to significant gravitational forces due to the masses of the moon and sun. These forces determine the motion of the earth relative to these two bodies. The gravitational forces on a particle at the surface of the earth will in general be different from those acting at the centre and the *difference* between these constitutes the tidal force at the surface particle.

Let us consider a single effective body of mass M situated at a distance R from the centre of the earth, and consider the tidal force on unit mass situated at a point P at the earth's surface, in line with the body, the radius of the earth being r ($\ll R$). The gravitational attraction on unit mass at the

6. THE TIDES

earth is $F = GM/R^2$ where G is the gravitational constant. Now on, reflection, it will be obvious that the tidal force at P will be

$$f = -r\frac{dF}{dR} = 2rGM/R^3.$$

This establishes the first important point, i.e. that the tidal force is directly proportional to the mass of the disturbing body, and inversely proportional to the *cube* of its distance. In consideration of the masses and distances of the sun and moon, it turns out that the tidal effect of the moon is just over twice as great as that of the sun. Thus the effect of the moon is dominant in tidal phenomena.

The distances of the sun and moon vary, the first annually and the second in a monthly cycle. Thus if we take the tidal force due to the moon in perigee as unity, the relative forces concerned may be summarized as in Table 1.

Table 1. Relative magnitude of tidal forces

	Perigee	Apogee
Moon	1	0·67
Sun	0·385	0·381

From the table we note that the effect of the moon varies a great deal from perigee to apogee, while that of the sun does not. This is, of course, due to the greater eccentricity of the moon's orbit.

Before considering in any detail the movements of the sun and moon, it seems best here to interpose some general remarks. The tidal forces at any point can be resolved into a vertical and horizontal component. Since there is no question of the sea surface coming into equilibrium with the static gravitational situation, we can for practical purposes ignore the vertical components of tidal force. It must instead be recognised that it is the horizontal component of the tidal force which is effective in producing motion of the sea water. Large tidal effects are only generated in the vicinity of the great oceans, and it is fundamental to realise that coastal tides are not directly produced by the tidal forces but result from oscillations of the great ocean water masses.

This point cannot be over-emphasised. The effect of the tidal forces is to set up seiche-like oscillations of the great oceans, of quite small amplitude. At the fringe of these oceans, the water surges over the continental shelves as a series of progressive waves, whose amplitude builds up where conditions are favourable. Thus the shape, and presumably orientation, of the Bristol Channel is particularly favourable to the progressive build-up of the amplitude of the tidal wave. This is, however, only an extreme case of a general rule. It will be found that there is a similar but less marked increase of tidal range as we move inwards in most of our major inlets. The speed of advance of the tidal wave is dependent on the depth, as the wavelength is obviously much greater than the depth, and the speed rule for waves in shallow water naturally applies i.e. $c = \sqrt{gh}$ where g is the acceleration of gravity and h the depth in metres. Thus for a depth of 100 m the wave speed is about 60 knots.

It is not practicable to calculate the tidal height for any specified place from knowledge of the tidal forces. Instead of this, what is done in preparing tide tables is to make a long series of observations at the required point, and try to isolate the timing and amplitude of the various harmonic components in the observed tidal curve. Most of these components have their counterpart in the tide producing forces, which *are*, of course, predictable, and thus forecast tables can be prepared. Naturally, the continental shelf system can distort the incoming tidal wave and produce additional harmonics. It is because of the non-applicability of simple theory that we have refrained from expanding in detail the derivation of the tidal forces in absolute terms.

Though theoretical prediction is not practicable, yet there is naturally a more or less close qualitive analogy between the tides in most places, and the variations in the tidal forces, and we shall now see how far it is possible to account for the features discussed in the earlier part of the chapter; it being understood that only the grosser features of the astronomy will be considered.

We shall consider first those features that are largely independent of the declinations of the moon and sun, (Fig. 1). If *the declinations are treated as*

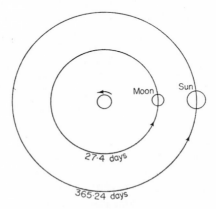

Fig. 1. Motion of the sun and moon relative to the earth and stars.

6. THE TIDES

zero and we look at the system from the north pole of the earth, then we see the rotation of the earth to be anticlockwise, and likewise the sun orbits in 365.24 solar days, and the moon orbits (also anticlockwise) in 27.4 days. During 1 solar day the sun advances by 1/365.24 of its orbit, and therefore the earth rotates relative to the fixed stars by 366.24/365.24 of a revolution.

Similarly, the moon, if aligned with the sun at a certain instant, will complete more than one orbit before it is again aligned. In fact, the time from alignment to re-alignment is 29.5 days, giving the ordinary lunar month, or period of repetition of the moon's phases. We can also say that the earth rotates *relative to the moon* in one lunar day of 29.5/28.5 of a solar day: that is 24.84 solar hr.

Thus since each body produces a tide raising force at both the proximal and remote end of the diameter aligned with it, we can see that there is a *solar* tide of period 12 hr, and a larger *lunar* tide of period 12.42 hr. We can usefully consider the resulting situation in two different ways, each of which contributes in its own way to comprehension of what is going on. Firstly we can think of the two tidal components beating together, and coming into phase with each other when the moon and sun are in line (either in the same direction—no moon, or in opposite directions—full moon). Thus, spring tides occur at no moon or full moon, whereas at the first and third quarters the two components are in antiphase, giving a neap tide. The tidal force at springs will therefore be the moon component plus the sun component, and at neaps will be the moon component minus the sun component.

If one spring tide happens when the moon is in perigee, the next will find the moon almost in apogee, thus we have our sequence of alternating large and small neaps. If perigee happens to occur at a neap tide we shall get a sequence of alternating large and small neaps, and our period of more or less equal ordinary spring tides. This was one of the features noted in the earlier part of the chapter.

If the moon is at perigee at a full moon, it will again be so after 385 days (moon's phase period being 29.5 and its orbital period 27.4 solar days). Thus if the moon is in perigee at a spring tide it will again be so after $192\frac{1}{2}$ days. Thus, the sequence of alternating spring tides will fade out and reappear after this period.

The alternative way to describe this particular aspect of the tidal forces is to think of a rotating vector force due to the moon, which is modified by the vectorial addition to it of the vector force due to the sun. We can then regard the tidal force as caused by the resultant vector (Fig. 2). In this figure we can consider a vector OC rotating clockwise at the uniform rate of the lunar semi-diurnal tide; i.e. completing one rotation in 12.42 hr. We may suppose OC to be proportional to the moon tide. If it be now spring tides we can suppose a vector *CS* corresponding to the sun tide, and directly supplementing the moon tide. Thus the resulting vector *OS* represents the tidal force on that occasion. After just over 7 days, the sun tide is represented by *CN* and the resultant tidal force by *ON*, this being the neap situation. At both springs and neaps, the resultant tide is *in phase* with the moon tide. However, after spring tides the resultant vector is in advance of the moon tide, this effect being maximal for the resultant *OS'*, when the sine of the lead angle θ is evidently (sun tide)/(moon tide) which is around 30°, giving a lead of resultant tide over moon tide of $12\cdot42 \times 30/360$ hr or about 1 hr. (Clearly the figure can be elaborated to take account of the eccentricity of the moon orbit, but this is not necessary for the explanation of the effect in question.)

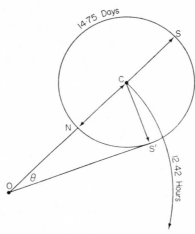

Fig. 2. Addition of the sun and moon vectors to give the resultant tide.

At this point the relationship between time of high water and tidal range, which we noted to be a universal rule, may be seen to be practically self-evident. Since the phase of the sun tide is obviously linked to time of day for any particular place, solar high water must always occur at some definite time, such as 03.00 and 15.00. At spring tides, the moon tide must be in phase with the sun tide, thus for that particular place spring tides will be linked to high water times around 03.00 and 15.00.

THE DIURNAL TIDE, DECLINATION EFFECTS

The above discussion, approximate in many ways, indicates that many of the obvious features of the semi-diurnal tide can be explained without reference to the declination of the sun or moon. We have recognized that the tidal force of the moon tends to form a high tide at each end of the diameter pointing to the moon. Suppose now the moon is in its highest North declination of around 28°, then the high water at the end of this diameter proximal to the moon will be situated in the northern hemisphere, and the high water at the distant end of the diameter will be situated in the southern hemisphere. Thus if we are ourselves observing in the northern hemisphere we shall encounter a higher high water followed by a lower high water. A more simple way of putting this is to say that, in any particular region, the moon tide has a diurnal component. This is true only if the moon's declination is non-zero. Similarly, if the sun's declination is non-zero, the sun-tide will have a diurnal component. A little thought will show that these diurnal components reinforce each other for either (a) no moon, both bodies in the same declination, or (b) full moon, sun and moon in opposite declinations. Thus the diurnal effects tend to maximise at periods of spring tides in summer and winter.

The moon's orbit is inclined by about 5° to the ecliptic, so the moon's maximum declination in certain years is about $23° + 5° = 28°$ while in other years the maximum declination is only $23° - 5° = 18°$. The situation of maximum declination, and therefore of maximum potential diurnal tides recurs every 18.6 years. Thus to obtain a full picture of the tidal variation at any point it is ideally desirable to continue systematic observations for a period of about 19 yr. It will be clear that further development of tidal theory in this elementary form cannot be taken further with profit and the above qualitative description can be regarded only as an introduction to this fascinating topic.

COASTAL TIDES

It has already been stated that tidal prediction is not in general possible from a purely theoretical consideration. The tidal forces produce oscillation in the great oceans which are not necessarily in phase with the forces themselves. As in any oscillating system there is a link between the natural frequency and the results of a periodic applied force. Standing waves may be set up and the open ocean tidal heights may lead or lag by indefinite amounts on the disturbing forces. Again, recapitulating, these oscillations spill over the continental shelves and long period waves are propagated over the shelves and into various semi-enclosed seas, to a degree dependent on their shape and orientation relative to the open ocean. The propagation rate is largely dependent on depth, and the progress of the tidal wave crest can be roughly followed by comparison of tidal gauges at various points. In the North Sea, for instance, charts are available showing the progression of the tidal wave and its height. The pattern is complex, since tidal waves enter in both north and south, and are reflected by the coast line. It is not surprising to find that nodal points are recognizable, where all effects cancel, and in fact the tidal pattern tends to rotate round these nodal points. These charts of co-tidal lines and tidal heights do not reduce well and the interested reader is recommended to obtain them from usual sources.

ACCURACY OF TIDAL PREDICTIONS

Because tide tables are based on long runs of information, and isolation of height and relative phase of each major component, they are calculated in an accurate way, and in important regions the predictions of high and low water heights are given to an accuracy of about 3 cm (0·1 foot). In addition, for major ports, typical tidal curves are given which allow interpolation for any time between high and low water to a similar accuracy. Don't believe them! The effects of barometric pressure and wind are not easy to predict, and generally it should be assumed that the errors of prediction are of the order of 20–30 cm. Very much larger discrepancies from forecast heights, of the order of 2–3 m, can occur in storm conditions. The subject is too specialized to pursue here but it may be said that the larger positive or negative surges originate, like the normal tides, in the oceans. The spurious wave so generated will typically follow a similar kind of track to the normal tidal wave, slowing in its progress and increasing in height as it moves from deeper to shallower regions.

As an example, to provoke thought, a barometric depression naturally tends to induce a hump in the sea surface. The static effect is evidently such that a local deficiency of 50 millibars would raise the local sea level by 50 cm but, of course, static situations are not to be found. Therefore, atmospheric pressure is quite significant in affecting sea level and furthermore, a depression moving at the natural speed of a long wave is likely to produce some measurable surge, and a depression which happens to follow the natural tidal path at the natural tidal speed, is likely to produce a large surge.

Thus, in the present state of knowledge, prediction of discrepancies from the tidal forecasts is perhaps even more important than improving the tidal forecasts themselves, and is as yet less well understood.

BIBLIOGRAPHY

Doodson, A. T. (1921). The harmonic development of the tide-generating potential. *Proc. Roy. Soc. A* **100**, 305–329.

Doodson, A. T. and Warburg, H. D. (1941). "Admiralty Manual of Tides", 270 pp. HMSO, London.

Chapter 7: Optics

The emission and absorbtion of light is a discrete process. Changes in the energy state of atoms can take place only by transition between possible energy levels, each of which corresponds to a non-radiating, steady state.

Between the emission of a discrete quantum of light and its absorption in some other element of matter, nothing observable takes place yet, for many historical reasons which will be well understood, various fictional entities have been created by the human mind to explain the process of light "transmission". Such are "rays of light" or "light waves", entities which though imaginary rather than observable, have proved to be powerful concepts in explaining image formation, refraction and the formation of diffraction patterns.

Since the speed of light c in free space is an accurately measured quantity, which we here take to be 3×10^8 m sec^{-1} and since the wavelength λ of light of any particular colour is also accurately known, we are easily led to infer a frequency v as characteristic of a given colour of radiation.

Now in all that has been said, we may extend the term "light" to cover all the well-known class of electromagnetic radiation e.g. infra-red, ultra-violet, X-ray and gamma radiation. All these radiations are characterised by the feature that the rest-mass of the particle (or quantum, or photon) concerned is zero. In this respect the radiation is different in character from radiations such as alpha or beta-radiation, or streams of neutrons, in which the rest-mass of the particle is non-zero, and where we may think of some actual particle of matter as being physically transported.

All these types of phenomena however, share in the wave characteristics notionally attributable to electromagnetic radiation. Both share in common the link between frequency of the wave character, and energy of the particle or quantum. This relationship is

$$\varepsilon = hv$$

where h is Plank's constant, and has a value of $6 \cdot 625 \times 10^{-34}$ J sec. It is necessary in considering the nature of any radiation to consider the flux ϕ which can be most conveniently expressed in *number* of particles or quanta per square metre per second, and also the *energy* per particle or quantum, which is most conveniently expressed in electron volts

$$1 \text{ eV} = 1\cdot 602 \times 10^{-19} \text{ J}.$$

The eV is also the unit used in describing the energy levels of atoms and is, of course, the energy required to move an electron through a potential difference of 1 V. Since most processes of emission or absorbtion of radiation consist in the displacement of an electron, or the ejection of an electron from its proper place in some molecular or atomic lattice, this selection of units is a consistent and suitable one.

We confine ourselves to electromagnetic radiation, in which the rest mass of a quantum is zero and the velocity in free space is c. This type of radiation is characterized also by the fact that it suffers no dispersion in free space: i.e. the group and phase velocities are identical and both equal to c.

In traversing matter, the phase velocity is in general less than c, and if the deduced phase velocity is v then we refer to the quantity $n = c/v$ as the "refractive index". A typical value for sea water is $n = 1\cdot 34$, thus the velocity of light in the sea may be taken as three-fourths of the freespace value for many practical purposes. Thus for example a camera looking through a plane window underwater must be focussed to three feet on its focussing scale, if objects at four feet are to be sharply defined.

Apart from the occasional introduction of artificial lighting for special purposes, the light under the sea surface is either daylight filtering downward from the sky, sun, moon or stars, or biological light produced by organisms. This biological light is an important feature of the deep sea, being the only justification for the preservation of the organs of sight in bathypelagic animals. Biological light is also a noticeable feature of the upper water at night. The whole topic is, however, more appropriate for a biological treatise than for a physical one, and will not be pursued here.

The light available at sea-surface level depends on the altitude of the sun, and on the cloudiness of the sky. At night the principal source of light is the moon.

There is an absence of published data covering the whole range of sky conditions, even for above-surface conditions. However, it will be obvious that the sea-surface will reflect light in accordance with the usual laws of optics, i.e. reflection will vary with the angle of incidence, and there will be some selection of the plane of polarization in the transmitted light.

It is not particularly easy to calculate the reflection accurately, as the sea surface is usually broken by waves and ripples. As a starting point, therefore, we may conveniently refer to Fig. 1, which gives the available light as observed immediately *below* the sea surface, as a function of the sun's actual (not apparent) altitude. This distinction is important at low altitudes, since the refraction of the atmosphere is then significant. The figure is expressed in photopic lux.

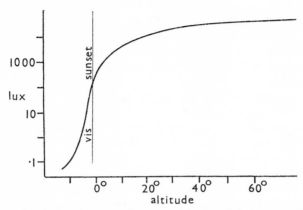

Fig. 1. Empirical relation between the true altitude of the sun, and the vertical blue–green illumination immediately below the sea surface. The curve refers to light overcast conditions. Visible sunset or rise occurs at a true altitude of minus 50 min approximately. Civil twilight and nautical twilight occur at true altitudes of $-6°$ and $-12°$ respectively. These twilight times are listed in the nautical almanack.

The spectral quality (photon energy distribution) for natural light varies considerably through full sunlight, overcast conditions and starlight conditions. Typical spectral curves are given in Fig.2.

The amount and spectral quality of available light within the sea depends of course on that available at the surface. But it also depends on the absorption of radiation by the sea water. The absorption, even within the visible spectrum, is markedly selective and varies both in amount and

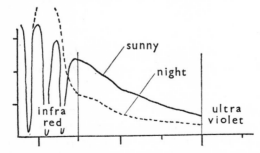

Fig. 2. Illustration of the spectrum at the surface in sunny conditions and at night when the effective colour temperature is lower. The measure is the relative number of photons at each energy level. (See scale of Fig. 4.)

spectral distribution from place to place in the sea. We may say that it depends on "water quality".

The range of variation is too wide and too inconsistent to be covered by any general statement, but a number of undoubted facts may be noted. (a) Inshore waters are more turbid than offshore. (b) Wave action or tidal action, conjoined to a muddy seabed, can give highly turbid conditions. (c) In the vicinity of rivers turbidity is generally increased; drastically so in times of flood.

We first consider some of the simple facts of the case. In ocean water, penetration of light is most effective in the blue–green region of the spectrum. As we approach inshore waters there is a tendency for the most effective penetration to shift into the green–yellow region. Taking blue–green light as our temporary standard, the depth at which available light has decreased to 1% is about 120 m in the oceans, decreasing in some shallow seas to around 30 m. In really turbid estuarine conditions values can occur more appropriately expressed in centimetres.

Because of this large range of variation it is necessary to be cautious of predictions unless they are based on observations in reasonable proximity to the area of interest. It is indeed desirable to *measure* the available light in all cases where a reasonably accurate knowledge of this factor is needed.

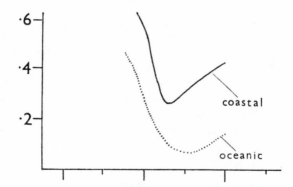

Fig. 3. Typical values of the diffuse attenuation coefficient per metre for a typical coastal and a typical oceanic water mass. (After Sverdrup *et al.*, 1942).

In Fig. 3 is shown the attenuation of light as a function of wavelength, for various conditions. As an example we may consider the implications of a curve such as that given for oceanic water. If we assume this attenuation is uniform with depth, we may calculate the available light at 50, 100, 200 and 400 m (Fig. 4). It will be seen that the spectrum is significantly restricted in comparison with that at the surface. The calculations are of course straightforward. If the attenuation coefficient is α, and the immediately sub-surface intensity in a restricted energy range is I_0, then we have at a depth z m that $I = I_0 e^{-\alpha z}$. Two important comments are needed.

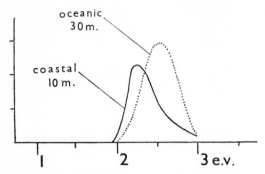

Fig. 4. We see here how, for the sunny spectrum at the surface, the attenuation has reduced the range of photon energies available at 10 m depth in coastal and at 30 m in oceanic water. Obviously at greater depths the available spectrum becomes more and more limited to the region of minimum attenuation.

Firstly, the examples are restricted to the visual part of the spectrum, since there is a lack of data on the other parts of the spectrum. Secondly, we might more properly refer to this attenuation coefficient as the *diffuse attenuation coefficient*.

By contrast we may consider the situation where a narrow beam of light is directed from a source within the medium onto a calibrated receptor. We may then consider the loss of light from the beam, and attempt to define another coefficient β, called the *beam attenuation coefficient*. While the diffuse attenuation depends on measurements which are straightforward, the absolute measurement of beam attenuation is less so. Ideally, one would pass a perfectly parallel beam from a light source through a plane window, through a definite path length in the sea, through another plane window onto a suitable, calibrated receptor.

To obtain zero attenuation as a standard, one would set up the equipment in air, then immerse in the water and note the reduction in light at the receptor. In principle, the only correction that must be made is to allow for the change in reflection at the two water–glass interfaces, as compared with the original air–glass interfaces. In practice, there are two other difficulties: first, that the beam may not be perfectly parallel light, so that refractive changes may alter the fraction of the beam influencing the receptor; secondly, that the temperature conditions may alter the brightness of the lamp on immersion.

The last named problem can be overcome, and must be overcome, by designing equipment which measures the *ratio* of the received light by the water path with that via a fixed path such as a fibre optic path. Only by such means can even comparative measures be made of beam attenuation. The difficulty of making these measures accurate in an absolute sense will be obvious, though of course accuracy will be increased as the length of the water path increases.

It is naturally tempting in these circumstances to calibrate the equipment by immersion in some known standard, such as distilled water. However, it is not easy to handle and maintain distilled water in a condition that renders it any better optically than many ocean waters. In practice, the standard may be more variable than the sample!

Emphasis has been placed on the practical difficulties of measurement in this one particular case, because accurate equipment for use in field work is largely unobtainable, and data correspondingly poor except in turbid sea areas, where the difficulty of measurement does not arise.

VISION UNDERWATER

We shall treat visual performance in a very general way to include the use of instruments, real or hypothetical, designed to form an image from the electromagnetic radiation. The situation differs from the terrestrial in three ways. (a) The beam attenuation coefficient is much greater than in air, the clearest water corresponding to a rather foggy day. (b) In most of the volume of the seas, light is severely limited, so that visual efficiency is controlled by quantum effects. (c) Because of the absorbtion of water, we are unable to have recourse to radiation in the infrared and microwave energy bands, which are proving powerful aids to night vision through the atmosphere.

RANGE OF VISION

The beam attenuation coefficient β has values ranging from about 0·15 per metre for the clearer oceanic regions, to indefinitely high values. The value is probably less than 0·30 for at least 90% of the seas and oceans of the world. It has been reported by Duntley (1963) that a dark body can in general be seen horizontally at a distance given by about $4/\beta$. Thus the range of vision in the clearer areas is 25–30 m and, in by far the greater part of the seas and oceans, it exceeds 15 m. However, the possibility of viewing at these, or any ranges, is dependent on the availability of sufficient light.

7. OPTICS

VISION IN LOW LIGHT

In this treatment we shall drastically simplify a topic which could if expanded fill many volumes. The aim is to bring out as clearly as possible the limitations which apply to any instrument, of which the animal eye is only one example, in viewing in low light.

We shall assume the laws of black body radiation as propounded by Max Planck. We shall assume the solar constant, and we shall build our approximations from these.

The power of black body radiation (Stefan's Law) from a square metre of surface is given by $W = 5.7 \times 10^{-8} T^4$ W, where T is the temperature in degrees Kelvin (absolute). The solar constant, which is the power received per square metre at a surface perpendicular to the sun's rays outside the atmosphere is 1350 W. The mean apparent diameter of the sun as viewed from the earth is 32' of arc.

To avoid any needless astronomical speculation, let us represent the sun by a spherical black body of 1 m radius—a model sun hung up in the sky. To have the same apparent size as the real sun, it would have to be viewed at a distance of 214 m. Using the expressions given above, we find that our model sun has to be at a temperature of 5,750°K to give the same radiation as the real sun, and its colour distribution is a good first approximation to that body's. To ge more precise, we note that the Planck law gives the number of photons to be

$$N_T = 1.5 \times 10^{15} T^3 \text{ m}^{-2} \text{ sec}^{-1}$$

and the energy distribution at any temperature of these photons is

$$N_\varepsilon = 10^{27} \frac{\varepsilon^2 \, d\varepsilon}{e^{\varepsilon/kt} - 1} \text{ photons m}^{-2} \text{ sec}^{-1}$$

where ε is in Ev, and k is Boltzmanns' constant and is equal to 86 electron microvolts per °K.

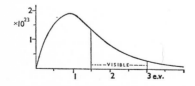

Fig. 5. The number of photons per second per electron millivolt energy interval emitted by one square metre of a black body at a temperature of 5,750°K. The total number of photons at this temperature is $2.90 \times 10^{26} \text{m}^{-2} \text{ sec}^{-1}$ and the total number at visible energies is 1.1×10^{26}.

If we calculate this distribution for a temperature of 5,750°K we obtain Fig. 5. For our present purposes we may assume that though heavy absorbtion occurs in other parts of the spectrum, the atmospheric absorption in clear conditions may be neglected in the main part of the visible spectrum. Thus we consider the relevant number of quanta as that in the visible energy range, $1\frac{1}{2}$ to 3 eV approximately. This is about $1\cdot1 \times 10^{26}$ photons m^{-2} sec^{-1} at the source, which is 1 m in radius. Applying the inverse square law we find that the number reaching a square metre of a plane perpendicular to the sun's rays at the earth is

$$1\cdot1 \times 10^{26}/214^2 = 2\cdot4 \times 10^{21}.$$

We shall now try to visualize a simple sphere which would be as bright to look at as a perfect diffuse reflector illuminated by this light. This can again be a 1 m radius sphere, but now emitting $2\cdot4 \times 10^{21}$ photons m^{-2} sec^{-1} in the visible energy range.

By extrapolation from recorded information (the reader may roughly check the reasonableness from Fig. 1) we know that this sphere is roughly as bright as a perfectly white surface illuminated by about 120,000 lux. Thus, if we reduced the illumination to 0·01 lux which is quite typical of the upper regions of the sea, the corresponding surface emission of our sphere would be 2×10^{14} photons m^{-2} sec^{-1}. If we now stand back at a distance of 214 m so that our sphere looks like an illuminated disc, the same apparent size as the sun, and, if we want to form an image in one twentieth of a second as we must if we are to see movement without blurring, then the number of photons per square metre available to us from the sphere is $(2 \times 10^{14})/(20 \times 214^2) = 2\cdot1 \times 10^8$ and if the aperture of our viewing lens is 8 mm, as it is for the dark adapted human eye, then the number of photons available to create an image is 14,000, and this will be true of any source of this brightness whose apparent size is that of the sun's disc.

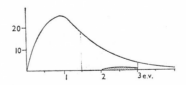

Fig. 6. This is the same frequency distribution as the previous figure, but scaled to give the number of photons per millivolt energy interval forming an image under the conditions described in the text. The hatched area is the product of this distribution with the quantum efficiency of an SbCs photocathode. The total number of photons in the visible is 14,000 and the number detectable as picture points by this cathode is about 1,000.

Thus if the receptor surface were fully efficient, over the whole energy range 1.5 to 3 eV there would be available 14,000 picture points to define the image. However, even a sensitive photocathode material such as Antimony–Caesium will only yield about 1,000 picture points in such a case (see Fig. 6) and if the receptor is the human retina, it is thought that around 200 picture points would be available. (Personal communication from Mr. R. L. Craig).

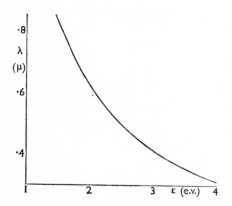

Fig. 7. The relation between λ in microns and ε in electron volts.

It will easily be seen that we are here approaching the limit where the insufficiency of photons dominates the quality of vision. If we become a stage more practical and suppose we are viewing a vertical white disc on a black ground, submerged to an ambient light level of 0·01 lux, we would lose a whole order of magnitude, and only about 100 picture points would be available from a sensitive photocathode. At 0·001 lux ambient only 10 points would be available which would not suffice to distinguish a disc from a square at this size.

It may be seen that there are definite limits set to the information that can be collected by a lens of given aperture. The efficiency of the eye cannot be improved even by an order of magnitude by the use of the best photocathodes.

The dependence of resolution directly upon the lens aperture leads to two fundamental conclusions.

1. The only way to gain improved vision is to increase the lens aperture. Thus a lens of 8 cm diameter will be 100 times more effective than one of 8 mm diameter. It will carry with it the disadvantage of a reduced depth of field.

2. Animals with smaller eyes will correspondingly be at a disadvantage, e.g. a fish with an entrance pupil of 0.8 mm would be 100 times less efficient. It follows that really small animals at these light levels cannot "see" in anything like the human meaning of the term.

It may be of some interest that large communities of active plankton animals, and pelagic fish, carry out vertical migrations, rising at night and descending by day, to an extent which roughly keeps them in light levels of the order we have been discussing above: i.e. light values around 0·01 to 0·001 as measured by a photometer directed vertically upward.

BIBLIOGRAPHY

Craig, R. E. and Craig, R. L. (1965). The prediction of undersea light, with special reference to Scottish fishing areas. *Photochem. Photobiol.* **4,** 633–636.
Duntley, S. Q. (1963). Light in the sea. *J. Opt. soc Am.* No. 1 **53,** 214–244.
Jerlov, N. G. (1968). "Optical Oceanography". 194 pp. Elsevier, Amsterdam.
Padgham, C. A. (1965). "Subjective Limitations on Physical Measurements". 78 pp. Chapman and Hall, London.
Utterback, C. L. and Jorgensen, W. (1936). Scattering of daylight in the sea. *J. Opt. soc. Am.* **26,** 257–263.

Chapter 8: Acoustics

Seawater is an excellent medium for the propagation of acoustic waves. Since it has serious limitations as an optical and radio medium, acoustics have played an important role in marine exploration.

The natural sounds in the sea have been studied principally because of the limitations they place on the use of acoustic instrumentation and to some extent, for their own sake. Apart from man-made sources, noise in the sea can be caused by wave action, sediment transport, and biological sources.

Since, as we will see, the attenuation of sound in the sea increases sharply with increasing frequency, the noise spectrum at any point usually shows the energy per Hertz to fall off with increasing frequency, at least at frequencies in excess of 1 kHz. This quieter background makes possible the use of high frequencies up to 1 MHz for short range exploration and study of the environment.

GENERAL PRINCIPLES

We first study and relate the physical parameters, pressure, particle velocity, and amplitude of vibration. For the general case, reference should be made to any textbook of sound. Here we shall consider only the simplest case of a *plane wave* proceeding in the x-positive direction. We assume that a progressive sinusoidal pressure wave can exist in the form:

$$p = P \cos \omega (t - x/c)$$

where, as in any wave motion, P is the peak pressure, and c is the phase velocity of the wavemotion.

To obtain the velocity u, the peak velocity U, and the amplitude A of the displacement, we may argue as follows: Consider a tube of unit cross-section area lying parallel to the x axis. Any element of thickness dx is subject to a force $-(\partial p/\partial x) dx$, and has a mass ρdx. Thus the acceleration of the element is $-(1/\rho)(\partial p/\partial x)$.

So
$$\frac{du}{dt} = -\frac{1}{\rho}\frac{\partial p}{\partial x}$$

or
$$u = -\frac{1}{\rho}\int \frac{\partial p}{\partial x}\,dt$$

$$= -\frac{P}{\rho}\int \frac{\partial}{\partial x}[\cos \omega(t - x/c)]\,dt$$

$$= -\frac{P}{\rho}\int \frac{\omega}{c}\sin \omega(t - x/c)\,dt$$

$$= \frac{P}{\rho c}\cos \omega(t - x/c)$$

$$= U \cos \omega(t - x/c),$$

that is the peak pressure and the peak velocity are related by

$$U = \frac{P}{\rho c}.$$

To find the amplitude of particle vibration we have, for any point x, that

$$u = U \cos \omega t$$

The amplitude is given by $A = \int u\,dt$ for a quarter cycle from maximum velocity to zero velocity.

This gives

$$A = \int_{\omega t = 0}^{\omega t = \pi/2} U \cos \omega t\,dt = U/\omega$$

so
$$A = P/\omega \rho c.$$

We note that in each of the expressions the quantity ρc appears as the parameter expressive of the properties of the medium. This product is given the name of "the characteristic acoustic impedance" of the medium, the analogy with electrical phenomena being thus recognised, with pressure as the analogue of voltage, and velocity as the analogue of current. In Table 1 the velocity and amplitude corresponding to a peak pressure of 1 pascal (1 N m^{-2}) is set out. In making the above calculation c is taken

Table 1.

Peak pressure 1 pascal or Peak velocity 0.65×10^{-6} m sec^{-1}.	
Frequency	Amplitude
1 H ($\omega = 2\pi$)	0.104×10^{-6} m
1 kHz	0.104×10^{-9} m
30 kHz	0.35×10^{-10} m
1 MHz	0.104×10^{-12} m

as 1500 m sec^{-1} (see Fig. 1), and ρ as 1027 kg m^{-3}, giving a value of ρc for sea water equal to 1.54×10^6 kg m^{-2} sec^{-1}.

The figure for 30 kH has been introduced since this is a typical frequency for sonar equipment. Since it is common practice to refer to a sound field by the rms rather than the peak value of the sound pressure, it may be helpful to restate the fundamental parameters on the basis of an rms pressure of 1 pascal.

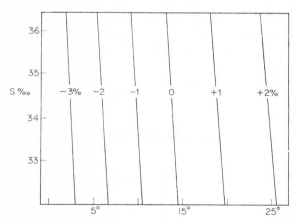

Fig. 1. The velocity of sound is taken in the text to be 1,500 m sec^{-1}. The exact value depends on temperature and salinity, and the percentage departure from the standard value is as above. An addition of 1% should also be made for each 800 metres of depth.

Table 2.

rms pressure 1 pascal or rms velocity 0.65×10^{-6} sec m^{-1}	
Frequency	Amplitude
1 Hz ($\omega = 2\pi$)	0.147×10^{-6} m
1 kHz	0.147×10^{-9} m
30 kHz	0.48×10^{-10} m
1 MHz	0.147×10^{-12} m

To exemplify a fairly high sound intensity, at a distance of 1 m from a sonar transducer (source level 100 decibels) working at 30 kHz, the amplitude would be 100 db above 0.48×10^{-10} m, i.e. would be 0.48×10^{-5} m or about 1/200 of a millimetre. If we consider a natural source such as a fish (taken to be a 10 db source) and we suppose the frequency to be 400 Hz, then the amplitude would be 0.11×10^{-8} m.

These examples are merely intended to put flesh and blood onto the theoretical bones. We proceed to consider the practical systems used in the measurement of sound fields, source levels and target levels.

PRACTICAL UNITS

The intensity of sound in the sea depends on the square of the sound pressure. The standard unit of intensity is that of a field with an rms sound pressure of 1 pascal. Observed sound intensities are described as plus or minus so many decibels with respect to 1 pascal rms.

Standard *source level* is such that it gives rise to a sound intensity of 1 pascal rms at a distance of 1 m, in the direction of greatest intensity. Thus we can describe any source level S in decibels with respect to a standard source.

In a large homogeneous body of water, the energy spreads uniformly, so the intensity at a range R from a source level S would be, in the absence of losses:

$$I = S - 20 \log R \text{ (logarithms to base 10)}.$$

There is however some attenuation due to the conversion of sound energy to heat within the medium (Fig. 2). Let this be α decibels per metre, then the

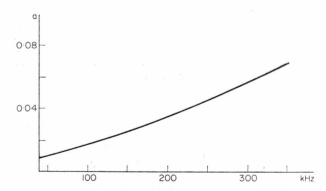

Fig. 2. The variation with frequency of the attenuation coefficient α (in decibels per metre). The figure is an approximation, based on somewhat conflicting reports.

corrected expression for sound intensity becomes:

$$I = S - 20 \log R - \alpha R.$$

This expression is only adequate for the assumed condition of uniform spreading in a large volume of water. Thus, it is reasonably accurate in respect of echo-sounding, but it will be in error for energy transmitted in a horizontal direction either where the depth is small compared to the range, so that spreading is limited by boundaries, or when refraction is significant, causing the sound to be channelled and so giving excess energy in some directions and a deficiency in others.

The final term to be defined is "target level" which again is conveniently defined in decibels relative to a standard. The incidence of acoustic energy on a target causes the target to act as a secondary source. If the incidence of standard intensity causes the target to act as a standard source, then that target is said to be of standard level, or a "0 dB" target.

If the range of the target from the equipment be R the target level T, and the intensity incident upon the target I_T, then the intensity of the echo signal at the receiver will be, as discussed

$$I_R = I_T + T - 20 \log R - \alpha R.$$

However we have already noted that

$$I_T = S - 20 \log R - \alpha R$$

and combining these pieces of information we obtain the fundamental sonar equation

$$I_R = S - 40 \log R - 2\alpha R + T. \tag{1}$$

This equation applies to small targets which subtend an angle at the equipment much less than the beamwidth. It necessarily refers only to targets on the axis of the transmitted beam, as we have defined source level by reference to the direction of maximum intensity. Where an extended target exists, large compared with the equipment beamwidth, a larger and larger section of the target becomes involved in reflection as the range increases.

One example of such a target is the sea bed which, being extensive, naturally fills the beamwidth entirely irrespective of range. It therefore gives rise to an echo intensity at the receiver which varies with range as $(-20 \log R - 2\alpha R)$ and whose accurate prediction depends on the nature of the seabed and the directional properties of the transmitter as well as its source level.

Even this treatment however is not fully adequate and it becomes necessary in a detailed study: (a) To know the target strength per square metre for the particular substrate, (b) To know how this varies with angle of incidence, and (c) How the radiated energy of the source varies with angle.

By performing an integration through the solid angle of interest, the total returned energy may be calculated. However the range to different elements of the seabed is not a constant, and it will be found that the different parts of the echo do not necessarily reinforce each other, unless the pulse length is very long. Thus, prediction of the nature of the seabed echo, or of that from any extended target, is not simple but requires knowledge of many parameters.

THE VOLUME SCATTERING COEFFICIENT

Another situation in which a "$20 \log R$" variation applies is in the assessment of targets too numerous to be individually resolved. The intensity of the received signal will be proportional to the *pulse volume* (Fig. 3). If we regard the acoustic beam as a cone of solid angle ω_e, the pulse volume would be $V = R^2 \omega_e \cdot \frac{1}{2} c\tau$ where τ is the pulse duration in seconds. The number of targets contributing to the echo would then be nV where n is the mean number per cubic metre at the range R.

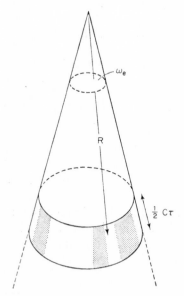

Fig. 3. The concepts of pulse volume and equivalent beam angle.

The intensity due to individual echoes sums, since the phase relationships are random. Therefore if each target is of level T decibels, the effective target strength for the pulse volume is T' where

$$T' = T + 10 \log n + 10 \log V$$
$$= T + 10 \log n + 20 \log R + 10 \log \omega_e + 10 \log (\tfrac{1}{2} c\tau)$$

and we write $S_v = T + 10 \log n$ we may regard S_v as the target strength per unit volume, or *volume scattering coefficient*.

By analogy with equation (1) we may write

$$I_r = S - 40 \log R - 2\alpha R + T'$$
$$= S - 20 \log R - 2\alpha R + [10 \log \omega_e + 10 \log (\tfrac{1}{2} c\tau)] + S_v \qquad (2)$$

where the square bracket encloses a parameter of the equipment which needs prior evaluation.

The term $10 \log (\tfrac{1}{2} c\tau)$ will be directly known, but the value of ω_e, the equivalent transducer beamwidth has to be determined by measurement of the beamshape of the transmit transducer used.

If the intensity on the axis is I_0, and the intensity at any other direction at the same range is I then the equivalent transducer beam angle will be

$$\omega_e = \frac{1}{I_0} \int I \, d\omega$$

where the integral is taken over the hemisphere.

Equation (2) may be taken to be the second sonar equation, as applying to multiple targets.

It should be noted that the sonar equations as stated, refer to the acoustic intensity I_r at the receive transducer. If we wish to use the electric signal from the receive transducer as a measure, we must take account of the directivity of that transducer in the calculations. For instance, if the same or an identical transducer is used for reception as is used to transmit, then we can assume an identical beamshape.

In that case we would have an equivalent *equipment* beam angle given by

$$\omega_e' = \frac{1}{I_0^2} \int I^2 \, d\omega.$$

POWER

It is now necessary to return to the concept of intensity which has, so far, been measured in decibels relative to 1 pascal and see how this relates to the other obvious description of intensity in Watts per square metre.

We again restrict ourselves to a plane wave, and we have already noted that in such a wave the pressure p and the particle velocity u are in phase, and can be represented at any plane perpendicular to our tube of unit cross-section area by

$$p = P \cos \omega t$$
$$u = U \cos \omega t$$

where $\quad U = P/\rho c.$

The instantaneous rate of work done at the plane is evidently

$$pu = \frac{P^2}{\rho c} \cos^2 \omega t$$

and the time mean of this is of course

$$W = \tfrac{1}{2} P^2/\rho c \quad \text{or} \quad P^2_{\text{rms}}/\rho c.$$

Thus an intensity of 0 dB relative to 1 pascal rms represents

$$\frac{1}{\rho c} \quad \text{i.e.} \quad \frac{1}{1.54 \times 10^6} = 0.65 \times 10^{-6} \text{ W m}^{-2}.$$

An omnidirectional source of acoustic power 1 watt would give rise to an intensity at 1 metre distance, of $1/(4\pi)$ W m^{-2}. This intensity exceeds that of 1 pascal rms by a factor of

$$\frac{1}{4\pi} \times 1.54 \times 10^6 = 1.23 \times 10^5.$$

The intensity would therefore be 51 decibels with respect to 1 pascal rms. and such a source would have a *source level* of 51 dB.

If a 1 W acoustic source were directional in its sound transmission, and were found to give an intensity of, say, 71 dB in its most efficient direction, it would be described as having a *directivity index* of 20 dB, by virtue of which it has acquired a source level of 71 dB. Thus we see again that source level depends on both power and directivity.

BIBLIOGRAPHY

Tavolga, W. N. (ed) (1964). "Marine Bio-Acoustics", 413 pp. Pergamon, New York.
Tucker, D. G. and Gazey, B. K. (1966). "Applied Underwater Acoustics", 244 pp. Pergamon, Oxford.
Urick, R. J. (1967). "Principles of Underwater Sound for Engineers", 342 pp. McGraw-Hill, New York.

Appendix: Some Properties of Sea Water and Other Constants

Chlorinity The basis of the present definition is a standard (Copenhagen Standard) seawater. The chlorinity of a sample is defined in terms of the ratio of its conductivity to that of standard seawater, at one of certain standard temperatures. For the greater part of the earth's water masses Cl‰ lies in the range 18 to 20‰. (The symbol ‰ represents parts per thousand by weight, generally referred to as "per mille".)

Salinity Defined as S‰ $= 0.03 + 1.805$ Cl‰ this definition implies that the salinity, which is approximately the concentration of all dissolved salts, has a finite value even when the halides are negligible, as in river water for instance.

Composition At a chlorinity of 19‰ the principal ions present are:

Chloride	18·98 ‰	Sodium	10·56 ‰
Sulphate	2·65	Magnesium	1·27
Bicarbonate	0·14	Calcium	0·40
Bromide	0·06	Potassium	0·38

Dissolved atmospheric gasses The saturation concentration of oxygen and nitrogen depend only slightly upon chlorinity, but are strongly temperature dependent. For Cl $= 19$‰ the saturation values in p.p.m. (parts per million by weight) are as follows:

	Oxygen	Nitrogen
0°C	8·08	14·40
10°C	6·44	11·56
20°C	5·38	9·65

Density Oceanographic tables are required to obtain the accuracies necessary for dynamic calculation. For approximate purposes the following tabulation by Helland–Hansen may be useful. It has been expressed in kilograms m^{-3} and refers to atmospheric pressure.

	$S = 25‰$	30‰	35‰	40‰
0°C	1020·09	1024·11	1028·13	1032·17
5°C	19·80	23·75	27·70	31·67
10°C	19·20	23·08	26·97	30·88
20°C	17·20	20·99	24·78	29·85

Density at great depths The effect of pressure is to increase the density, which for 0°C and 35‰ salinity varies as below:

Depth	Density (kg m^{-3})
0	1028·13
2,000 m	37·47
4,000 m	45·35
6,000 m	55·35

Conductivity The relation between salinity and conductivity is obtainable to high accuracy using oceanographic tables. The following will suffice for some purposes. It is expressed in ohms^{-1} m and refers to atmospheric pressure.

	$S = 25‰$	30‰	35‰	40‰
0°C	2·1	2·5	2·9	3·2
10°C	2·8	3·3	3·9	4·3
20°C	3·5	4·2	4·9	5·4

Refractive index

	$S = 25‰$	30‰	35‰	40‰
0°C	1·3390	1·3398	1·3407	1·3419
10°C	3384	3393	3401	3412
20°C	3375	3384	3393	3402

Freezing point

$S = 10‰$	20‰	30‰	35‰	40‰
−0·53°C	−1·07°C	−1·63°C	−1·91°C	−2·20°C

Angular velocity of earth $\omega = 73 \times 10^{-6}$ radians sec^{-1}

Solar constant $W = 1350$ W m^{-2}.

Index

Acoustic
  amplitude, 74
  attenuation, 76
  intensity, 76, 79
  pressure, 74

Beam attenuation, 69
Bernoulli, 50
Black body, 69

Craig, R. L., 71, 72
Coriolis force, 3, 17

Decibar, 2
Diffuse attenuation, 66, 67
Directivity index, 80
Duntley, S. Q., 68, 72

Eddy conductivity, 12
  diffusion, 9
  viscosity, 12, 15, 24
Ekman, V. W., 34
Estuary, 41

Gade, H. G., 10, 16

Ice, 43

Lens aperture, 71
Light
  quantum theory, 63, 69
  sub-surface, 66

Mixed layer, 14, 30

North Sea, 7, 8, 55
Nutrients, 14

Oslofiord, 10, 16

Plane waves, 73
Planck, Max, 69
  constant, 63, 69

Slope due wind, 36
Solar constant, 69, 82
Sonar equation, 77, 79
Source level, 76, 80
Spiral of velocities, 27
Stress, viscous, 15, 24, 27
  wind, 23, 27, 33, 35, 36

Tait, J. B., 6, 8
Target level, 77
Thermocline, 9, 38
Tidal force, 56
Tides, British, 56
  diurnal, 60
  semi-diurnal, 57
Transport, due to wind, 27, 36
Tulloch, D. S., 6, 8

Velocity of sound, 75
Viscous dissipation, 31
Vision, and turbidity, 68
  and available light, 69

Waves, cycloidal, 44
  decrease with depth, 47
  deep water, 44
  group velocity, 54
  phase velocity, 45
  shallow water, 50
  short crested, 44